Oxford Physics Seri

MW01025802

General Editors

E. J. BURGE D. J. E. INGRAM J. A. D. MATTHEW

Oxford Physics Series

1. F. N. H. Robinson: *Electromagnetism*
2. G. Lancaster: *D.c. and a.c. circuits. Second edition.*
3. D. J. E. Ingram: *Radiation and quantum physics*
5. B. R. Jennings and V. J. Morris: *Atoms in contact*
7. R. L. F. Boyd: *Space physics, the study of plasmas in space*
8. J. L. Martin: *Basic quantum mechanics*
9. H. M. Rosenberg: *The solid state. Second edition.*
10. J. G. Taylor: *Special relativity*
11. M. Prutton: *Surface physics*
12. G. A. Jones: *The properties of nuclei*
13. E. J. Burge: *Atomic nuclei and their particles*
14. W. T. Welford: *Optics. Second edition.*
15. M. Rowan-Robinson: *Cosmology. Second edition.*
16. D. A. Fraser: *The physics of semiconductor devices. Second edition.*
17. L. Mackinnon. *Mechanics and motion*

J. L. MARTIN

KINGS COLLEGE, LONDON

Basic quantum mechanics

Clarendon Press · Oxford · 1981

Oxford University Press, Walton Street, Oxford OX2 6DP

OXFORD LONDON GLASGOW
NEW YORK TORONTO MELBOURNE WELLINGTON
KUALA LUMPUR SINGAPORE JAKARTA HONG KONG TOKYO
DELHI BOMBAY CALCUTTA MADRAS KARACHI
NAIROBI DAR ES SALAAM CAPE TOWN

Published in the United States by
Oxford University Press, New York.

British Library Cataloguing in Publication Data

Martin, John Legat
Basic quantum mechanics.—(Oxford physics series).
1. Quantum theory
I. Title
530.1′2 QC174.12 80-41845

ISBN 0-19-851815-3
ISBN 0-19-851816-1 Pbk

Typeset by Eta Services, Beccles, Suffolk.
Printed in Great Britain by Billing & Sons Ltd.
Guildford London Oxford Worcester

Preface

I cannot believe that God plays dice with the cosmos.

Albert Einstein.

We throw the dice. It is up to the Lord to fix how they are to fall.

Proverbs 16:33.

THE basic rules of quantum mechanics are not at all complicated: the main problem is the hurdle of unfamiliarity. We have no right to expect that 'everyday' experience will be of any help in those regions of physics which cannot be handled in an everyday way. This is easy to accept. Yet it has taken centuries to learn this lesson, and it has only been relatively recently that physicists have been forced to accept that 'everyday' mechanics must be modified to deal with phenomena on the cosmological or on the atomic scale. In the main, quantum mechanics is concerned with the atomic scale. Moreover, quantum mechanics is *statistical* in nature: it deals with questions of probability. This is in itself a psychological barrier for many people: even Albert Einstein was inflexibly opposed to a theoretical scheme which leaves so much to chance.

The basic rules are laid down in the first three chapters and repeatedly illustrated by reference to one and only one example: the states of polarization of a single photon. At this stage the mathematics is kept as simple as it can be, since it is important in a subject where conceptual difficulties may be severe not to make matters worse!

Understanding the first three chapters is not the same as being able to *use* quantum mechanics. The rules may be simple, but the range of applications is very rich, and the solutions of some problems lie well beyond the reach of present mathematical expertise. In order to be a competent quantum 'mechanic', one must become familiar with techniques appropriate to more general systems, and understand why classical ideas like energy and momentum reappear in quantum mechanics. Such topics, along with their applications in particular to electron spin, the hydrogen atom, and the harmonic oscillator, take us up to chapter 11.

Very few realistic quantum problems are exactly solvable, and the next six chapters deal with the more frequently used approximation methods. Throughout I have tried to offer realistic applications of each method, to show that they all have a genuine value. The book ends with a brief sketch of a few of the many difficulties and problems which have plagued the quantum philosophers from the very beginning.

Though the text originates from two undergraduate courses given at King's College, it is not a mere expansion of the lecture notes; most topics are presented in much greater depth. The emphasis throughout is on a careful and precise statement of quantum principles, with the mathematics no more complicated than is required for this aim. I have not followed the historical development of the subject, fascinating though it is, as it covers 25 years of groping and false starts. I feel that the best time to meet, for example, Bohr's 1913 model of the hydrogen atom is *after* meeting Schrodinger's 1925 model. In this way, one may appreciate much more fully both the brilliance and the shortcomings of Bohr's approach.

There are problems scattered through the text to encourage the reader to test his grasp of the subject reasonably frequently. Sometimes the result of a problem is used at a later point in the text; this should not cause any difficulty. When there is a need for techniques which can be found in any of the dozens of books on mathematical methods, they are not reproduced; instead they are summarized as lists of mathematical prerequisites.

It is obvious that the author of a book such as this must owe a great deal to others. In particular I wish to acknowledge both the helpful discussions with several colleagues, especially Professor E. J. Burge, and the very many texts on quantum mechanics which already exist: these are so numerous that it would be impracticable to list them all, and invidious to list only some. I am grateful to the Editors of *Physics Today* for permission to include the quotations on pp. 103 and 122. Several of the diagrams, mainly in chapters 16 to 18, were generated by FORTRAN programs invoking the very useful DIMFILM routines available at the University of London Computer Centre.

King's College, London.
April 1981

J.L.M.

Contents

1. Basics

Why do we need quantum mechanics?

IT is now about three hundred years since Newton laid the foundations of his mechanics, and rather more than a hundred since Maxwell formulated the equations of electromagnetic theory. Up till about the end of the nineteenth century it was generally believed that Newtonian mechanics and Maxwell's equations were the complete and final expression of the basic laws of physics, and that future progress lay merely in minor refinements and more exact solutions. This belief was not unreasonable, since the success of Newtonian mechanics had been so spectacular till then. It had even been said by some that there was nothing left to be discovered by the next generation! However, there were misgivings; in connection with an apparently insurmountable difficulty in black-body radiation, Lord Kelvin referred in 1904 to 'a cloud which has obscured the brilliance of the molecular theory of heat and light during the last quarter of the nineteenth century'; and indeed it was becoming clear that certain awkward experimental results would compel a major theoretical change. It was not till 1925 that the final shape of the 'new' dynamics was found; since then the basic ideas have not changed.

From our present point of view, perhaps the most important feature of Newtonian mechanics is that it is *deterministic*. If we know every detail of the state of a physical system at a particular time, then we are able in principle to predict the precise result of an observation made at a later time. However, to know whether the physical world is indeed deterministic we must make an appeal to experiment, and experiment at the atomic level apparently suggests that the physical world is *not* deterministic (at least, not in the sense just described). Thus Newtonian mechanics needs to be modified.

Why then has Newtonian mechanics been so successful? An important feature of quantum mechanics is that certain fundamental physical quantities (previously believed to be continuous) are observed to take only a discrete set of values. For example, angular momentum will take a value which is an integer multiple of a basic unit, $\frac{1}{2}\hbar = 0{\cdot}527 \times 10^{-34}$ joule-seconds (J s), and no other value. This 'natural' unit is almost unimaginably small compared with 'everyday' angular momenta, and it is not surprising that it was overlooked. Generally speaking, it is correct to use Newtonian mechanics for systems big enough or massive enough for quantum effects to be negligible, e.g. a railway train, or even a particle large enough to be visible in an optical microscope.

Will quantum mechanics survive, or will it be modified as Newtonian mechanics has been? No-one knows, but the history of science suggests that we would do well not to accept any theory as necessarily final, however beautiful or however reasonable it may seem.

The photoelectric effect

A good example of a phenomenon inexplicable by classical theories is the *photoelectric effect*. In 1888, Hallwachs found that a negatively charged metal plate could be discharged by ultraviolet light, and it was soon understood that this results from the ejection of electrons from the surface of the metal. Ten years later, Lenard investigated the phenomenon, obtaining results that were incomprehensible to the theory of the time. He showed that for monochromatic light of frequency ν the kinetic energy of an ejected electron could not exceed a maximum which depended linearly on ν:

$$\text{K.E.} \leqslant h\nu - E_0;$$

h was apparently a universal constant† ($\sim 6{\cdot}6 \times 10^{-34}$ J s), while E_0 depended on the metal being irradiated. Classically, one would have expected that the maximum kinetic energy would increase with the incident intensity, but this did not happen; perhaps more surprisingly, for radiation whose frequency was too low ($h\nu < E_0$) no electrons at all were ejected, *however intense* the radiation.

In 1905 Einstein gave a quantum explanation of this phenomenon which still stands today. The monochromatic beam is in this situation to be regarded as a beam of *photons* each carrying the same characteristic amount of energy $h\nu$. When a photon in the beam interacts with an electron in the metal, the electron may be ejected with kinetic energy not greater than $h\nu - E_0$, where E_0 is the minimum energy required to overcome the potential 'cliff' at the surface. The photon description fits the observed effects exactly. Any classical attempt at a description fails since it cannot allow for the discrete nature of the energy in a monochromatic beam.

Plane-polarized light

A useful illustration which we shall use repeatedly is provided by the properties of polarized light. Imagine a uniform monochromatic collimated beam of light. Such a beam has definite frequency, direction, and intensity, but is not quite completely specified; there remains the possibility of polarization. We shall consider plane-polarization to begin with.

A beam is *plane-polarized* if the electric vector $\mathscr{E}$ in the beam is everywhere parallel to some vector $\boldsymbol{a}$, which must itself be perpendicular to the direction of propagation. The direction of $\boldsymbol{a}$ is conveniently specified by the angle θ between $\boldsymbol{a}$ and a fixed vector $\boldsymbol{a}_0$; we shall call the corresponding polarization P_θ. There is one such polarization P_θ for each θ in the range $0 \leqslant \theta < \pi$; note that P_θ and $\mathrm{P}_{\theta+\pi}$ are the same polarization, since the *sense* of the vector $\boldsymbol{a}$ is not relevant.

† **Planck's constant.** Introduced by Planck (1900) for reasons from *statistical* mechanics, this constant appears throughout the entire range of microscopic physics.

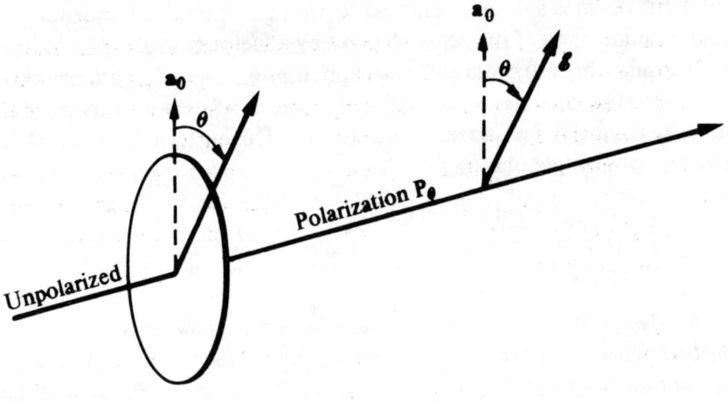

One way of obtaining a beam polarized at the angle θ is to pass it through an appropriately oriented plane-polarizer (such as a Nicol prism or a Polaroid filter). By an obvious convention we say that a plane-polarizer which produces the polarization P_θ is oriented at the angle θ.

Plane-polarized photons

Phenomena like the photoelectric effect lead us to regard a beam of P_θ-polarized light as a beam of photons, and we then say that each photon in the beam is 'in the **state** P_θ' (see below, p. 6).† Quantum mechanics is the study of the relations between such states.

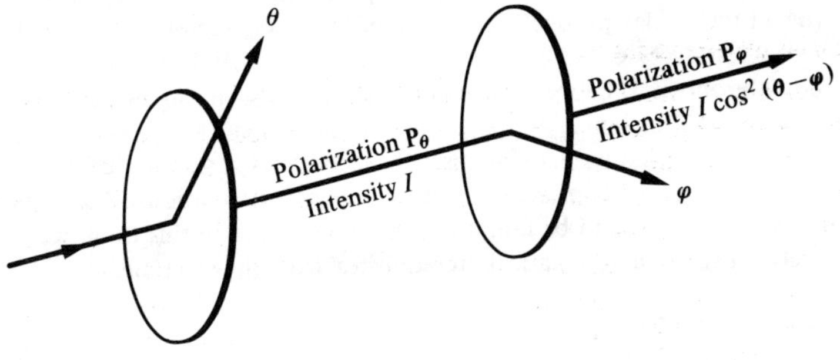

Consider the following known experimental facts: a monochromatic beam with polarization P_θ encounters a polarizer set at angle ϕ; then

(1) the beam exits with polarization P_ϕ (as we already know);

† In the text bold type will always indicate the point at which a *definition* of an essential concept is given. This point need not be where the concept is first mentioned.

(2) the beam is still monochromatic, with no change in frequency;
(3) the intensity of the beam changes by a factor $\cos^2(\theta-\phi)$.

Fact (2) implies that the beam consists of photons of energy $h\nu$ both before and after meeting the polarizer; then fact (3) can only imply that the *proportion* of photons surviving the polarizer is $\cos^2(\theta-\phi)$. This in turn suggests that we should say for a single photon

$$\left.\begin{array}{l}\text{the } \textbf{probability} \text{ that a photon in state } P_\theta \\ \text{survives a polarizer with orientation } \phi \\ \text{(thus then being in state } P_\phi)\end{array}\right\} = \cos^2(\theta-\phi)$$

Here, as always in quantum mechanics, the word *probability* really means *proportion*, where we have a large number of identical situations in mind, e.g. a large number of photons striking a polarizer. It should now be clear that it is not easy to avoid the idea of indeterminacy when we wish to describe a beam of light as a stream of photons; it is usually not possible to say with certainty what will happen to any particular photon in the beam.

A physical theory must relate and predict the results of experiments. Thus quantum mechanics must be able to deal with probabilities. We are given a clue to the appropriate formalism by noting that

$$\cos(\theta-\phi) = \cos\theta\cos\phi+\sin\theta\sin\phi, \tag{1.1}$$

suggesting that we should associate number pairs ('vectors') with the polarization states

$$P_\theta \leftarrow (\cos\theta, \sin\theta), \qquad P_\phi \leftarrow (\cos\phi, \sin\phi).$$

('P_θ is *represented* by $(\cos\theta, \sin\theta)$.') The probability may now be written in terms of the 'scalar product' (eqn (1.1)) of these two vectors. This is not an accident: in every case where quantum mechanics has been used the link, indeed the only link, between theory and experiment is through scalar product expressions of this kind.

Quantum mechanics therefore uses the techniques of vector algebra, along with extensions appropriate to more complicated situations. Accounts of these techniques are to be found in other books, and in this book we shall merely list some 'mathematical prerequisites' from time to time.

Mathematical prerequisites

An n-component **column vector** is an ordered set of n complex **components**, conventionally arranged as a vertical array

$$c = \begin{pmatrix} c_1 \\ c_2 \\ \vdots \\ c_n \end{pmatrix}.$$

The **Hermitian conjugate** of a column vector is a **row vector**, whose components are the complex conjugates c_i^* of those of the column vector, arranged conventionally as a horizontal array:

$$c^+ = (c_1^*, c_2^*, \ldots, c_n^*)$$

(often called 'c-dagger').

The **sum** of two column vectors is obtained by adding corresponding components:

$$\begin{pmatrix} a_1 \\ a_2 \\ \vdots \\ a_n \end{pmatrix} + \begin{pmatrix} b_1 \\ b_2 \\ \vdots \\ b_n \end{pmatrix} = \begin{pmatrix} a_1+b_1 \\ a_2+b_2 \\ \vdots \\ a_n+b_n \end{pmatrix}.$$

The sum of two row vectors is similarly defined. We may multiply a vector by a (possibly complex) number, according to

$$\lambda \begin{pmatrix} c_1 \\ c_2 \\ \vdots \\ c_n \end{pmatrix} = \begin{pmatrix} \lambda c_1 \\ \lambda c_2 \\ \vdots \\ \lambda c_n \end{pmatrix}.$$

The **zero vector** (either row or column) has all components zero and is usually written 0.

The **scalar product** of a row vector w^+ and a column vector v is

$$w^+ v \equiv (w_1^*, w_2^*, \ldots, w_n^*) \begin{pmatrix} v_1 \\ v_2 \\ \vdots \\ v_n \end{pmatrix} = \sum_{i=1}^{n} w_i^* v_i.$$

If $w^+ v = 0$, the column vectors v and w (or equivalently the row vectors v^+ and w^+) are **orthogonal**.

The value of $v^+ v$ for any vector v is always positive, since

$$v^+ v = \sum_{i=1}^{n} |v_i|^2,$$

and thus is a sum of positive terms. The **length** of the column vector v is the positive root $(v^+ v)^{\frac{1}{2}}$. A vector is **normalized** if its length is 1.

The nature of a quantum state

We are now ready to look at the basic structure of quantum mechanics. A difficulty which we meet at the very beginning is that the ideas are unfamiliar. We are used to Newtonian mechanics where everything is *predictable*. For example, if we have to do with the motion of a particle under the influence of a given force, and if we are given the position and the velocity of that particle at time $t = 0$, it is 'simply' a matter of solving the Newtonian equations of motion to obtain the precise position (or the velocity) at any future time—or, for that matter, any past time; it works both ways. ('Simply' in quotes, because there may be practical numerical difficulties in solving the equations; such difficulties are not really relevant here.)

Our discussion of photon polarization, on the other hand, suggests that we are wrong to demand complete predictability in physics. To understand what a radical change in thinking is required, consider some 'quantum' system at time $t = 0$, and imagine that we have at that time subjected it to an observation O_1 so comprehensive that we have squeezed out every last drop of information about its state that we can. It is then natural to say that it is then in a *definite* **state** S which is specified by the information that we have extracted. We have not, as yet, introduced anything new; after all, if we have measured the position and velocity of a Newtonian particle (which we understand) we are equally entitled to say that it is then in a definite state.

The major change is that even though we declare that a *quantum* system is in a definite state S, a different observation O_2 does not necessarily yield a definite *result*. Sometimes we may get the result *a*, say, and sometimes *b*. Worse still, this same observation, when performed for a *different* state S′, may sometimes give the result *a* and sometimes *b*. In other words, knowing the result of an observation *cannot* tell us the state that the system was in when the observation was made; and knowing the state of a system *cannot* predict with certainty what the result of any observation will be.

In such a situation it is difficult to see how one can have a theory describing the mechanics of a *single* quantum system; in fact, no-one has ever succeeded in producing a satisfactory one. The claims of quantum mechanics are much more modest, however, and can be summarized as follows.

1. A thoroughly comprehensive observation O_1 of a system will put that system in a definite quantum state S; the particular state S which results is defined by the result of the observation. (Thus in one sense, the state S is a kind of *codification* of the result of the observation.)
2. A further observation O_2—possibly a different one and not necessarily comprehensive—made on a quantum system in a definite state S may produce one of a set of possible results $a_1, a_2, \ldots$. If there is more than one possible result, *no prediction is made* about which result will occur.
3. Now imagine that we use the comprehensive observation O_1 to put the system *repeatedly* into a definite quantum state; on those occasions when the quantum state is S (as checked by the result of O_1), let us then go on to make the further observation O_2. Quantum mechanics predicts the *proportion* p_1 *of occasions* for which the result is a_1, the *proportion* p_2 *of occasions* for which the result is a_2, and so on.

We may therefore say that quantum mechanics is concerned with the *statistical* outcome of a fixed experimental procedure repeated many times; its aim is to predict the values of the *probabilities* $p_1, p_2, \ldots$. It achieves this aim with the aid of an appropriate mathematical structure, which we must now consider.

The representation of states by state vectors

Quantum systems may be quite simple, or they may be extremely complicated; therefore the same will be true of the mathematical structures we set up to represent them. To begin with we consider the very simplest systems; certain generalizations will become necessary for the more complicated systems of Chapter 4.

For the simplest systems, quantum mechanics asserts that each possible physical state is to be **represented** by a normalized column vector (the **state vector**) or (equivalently) by the Hermitian conjugate row vector

$$\text{physical state} \leftarrow \begin{pmatrix} c_1 \\ c_2 \\ \vdots \\ c_n \end{pmatrix} \quad \text{or} \quad (c_1^*, c_2^*, \ldots, c_n^*) \qquad \text{(Rule I)}$$

with $\sum |c_k|^2 = 1$. The components c_k are to be allowed to be complex. How many components are needed depends on how complicated the system is; we have already hinted that photon polarization requires two (p. 4). Indeed, some important systems cannot be represented by state vectors with a finite number of components; they will not be considered until Chapter 4.

To complete the scheme of quantum mechanics we shall need some further rules to link the values of the components with the results of observation (see below), and others to provide a *dynamics*, i.e. equations of motion governing the change in the components as time passes (see Chapter 3).

The physical interpretation of the formalism

Let us begin with an illustration. Consider the simple *analyser* consisting of a plane-polarizer in front of a photon detector; the detector gives a signal ('yes') whenever a photon passes through the polarizer and meets the detector. An

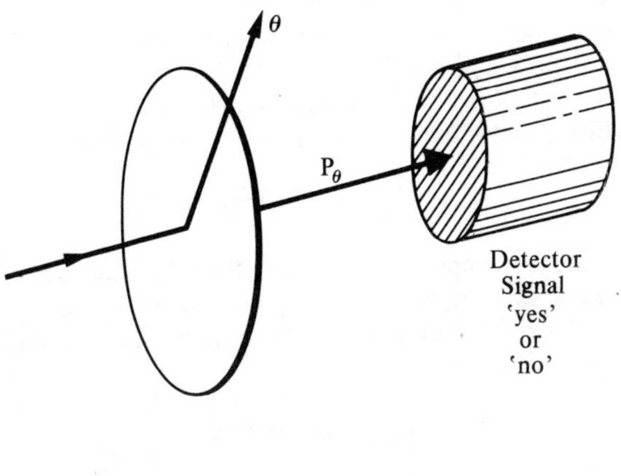

observation consists in aiming a photon at the analyser and sensing a signal ('yes') or the absence of a signal ('no'). When the analyser is oriented at angle θ, any photon in polarization state P_θ will certainly reach the detector and yield a 'yes'; as it happens, this is true for no other polarization state whatever. (Other states, of course, *may* give a 'yes', but not with certainty; the state P_ϕ will give 'yes' with probability $\cos^2(\theta-\phi)$.)

It should be clear that a measurement of this kind cannot determine the state of the incident photon. In fact, it is generally true in quantum mechanics that we never 'measure' states, but physical quantities; knowledge of the state implies knowledge of the probabilities of all the various possible outcomes, according to the rules now to be outlined.

To help us in this we introduce the idea of a **'simple' observation**, whose two possible outcomes are 'yes' and 'no', and for which there is *exactly one* state (say S) yielding 'yes' with certainty; let us call this observation O(S), labelling it with the state S in a natural way. Observations of this kind are not as artificial as they may seem; the photon detector just described provides the example of the 'simple' observation $O(P_\theta)$. In any case, more complicated observations are easily expressed in terms of 'simple' observations (Chapter 2), so that the idea is in no way restrictive. By the way, the 'comprehensive' observations mentioned earlier (p. 6) are 'simple' observations in this sense; they may *not* be simple in any practical experimental sense!

The fundamental connection between theory and observation can now be given. Imagine that a system in a state represented by the state vector w is subjected to the observation O(S), where the state S is represented by the state vector v. Then

$$\left.\begin{array}{l}\text{the probability that the outcome}\\ \text{of the observation is 'yes'}\end{array}\right\}\ \text{is } |w^+v|^2. \qquad \text{(Rule II)}$$

It is important to understand that this rule provides the *only* connection between the formalism and the physics. It expresses the physical relation between states in terms of a scalar product, and it is for this reason that quantum mechanics uses the methods of linear algebra—the natural mathematical apparatus for dealing with scalar products.

Let us see how Rules I and II work out for a plane-polarized photon. We have already hinted (p. 4) that the state P_θ may be represented by

$$P_\theta \leftarrow \begin{pmatrix}\cos\theta\\ \sin\theta\end{pmatrix} \quad \text{or} \quad (\cos\theta, \sin\theta).$$

Let us perform the 'simple' observation $O(P_\phi)$ on a photon in the state P_θ (or, in everyday language, allow a plane-polarized photon, state P_θ, to fall on the analyser described above, oriented at angle ϕ). Applying Rule II in this case leads us to write down

$$w^+ = (\cos\phi, \sin\phi) \quad \text{and} \quad v = \begin{pmatrix}\cos\theta\\ \sin\theta\end{pmatrix},$$

and then to write

$$\left.\begin{array}{r}\text{the probability of getting}\\ \text{a signal from the detector}\end{array}\right\}\text{ is }\left|(\cos\phi, \sin\phi)\begin{pmatrix}\cos\theta\\ \sin\theta\end{pmatrix}\right|^2$$

$$= |\cos\phi\cos\theta+\sin\phi\sin\theta|^2$$

$$= \cos^2(\theta-\phi).$$

So the rules work for a plane-polarized photon, using a state vector with two components.

PROBLEMS

1. Use Rule II to confirm that making the observation O(S) on the state S gives 'yes' with certainty. (Recall that state vectors are required to be normalized.)

2. u and v are any two fixed normalized vectors, and α is a complex number.
   (a) Prove that $(u^+ + \alpha^* v^+)(u+\alpha v) \geqslant 0$ for any α. (The left side is a sum of squared moduli.)
   (b) Prove that as α varies, the minimum possible value of $(u^+ + \alpha^* v^+)(u+\alpha v)$ is $1-|u^+v|^2$, attained when $\alpha = -v^+u$. (One route to this result is to remark that the partial derivatives with respect to the real and imaginary parts of α must both vanish at the minimum. There is a short cut: differentiate with respect to α^*, keeping α 'constant'.)
   (c) Deduce that $0 \leqslant |u^+v|^2 \leqslant 1$, and thus confirm that the scalar product expression in Rule II may safely be interpreted as a probability.

3. Suppose that the state S_k is represented by the vector

$$S_k \leftarrow \begin{pmatrix}0\\0\\ \vdots\\0\\1\\0\\ \vdots\\0\end{pmatrix},$$

with all components zero, except $c_k = 1$. For a *general* state

$$S \leftarrow \begin{pmatrix}c_1\\c_2\\ \vdots\\c_k\\ \vdots\\c_n\end{pmatrix},$$

show that the probability that the observation $O(S_k)$ yields a 'yes' is $|c_k|^2$. (The components c_k are for this reason often called **probability amplitudes** in the literature. However, 'amplitude' is a much overworked word, and we shall not use it.)

Representations are not unique!

The representation of a given quantum system is never unique. For example, we could just as well have taken

$$P_\theta \leftarrow \frac{1}{\sqrt{2}}\begin{pmatrix} e^{i\theta} \\ e^{-i\theta} \end{pmatrix} \quad \text{or} \quad \frac{1}{\sqrt{2}}(e^{-i\theta}, e^{i\theta}).$$

(Remember that the elements have to be complex-conjugated in the row vector.) If we consider the same physical situation as before, where a photon in state P_θ falls on an analyser oriented at ϕ, Rule II leads us to write

$$w^+ = \frac{1}{\sqrt{2}}(e^{-i\phi}, e^{i\phi}) \quad \text{and} \quad v = \frac{1}{\sqrt{2}}\begin{pmatrix} e^{i\theta} \\ e^{-i\theta} \end{pmatrix},$$

and then

$$\left.\begin{matrix}\text{the probability of getting} \\ \text{a signal from the detector}\end{matrix}\right\} \text{is} \left| \frac{1}{\sqrt{2}}(e^{-i\phi}, e^{i\phi}) \frac{1}{\sqrt{2}}\begin{pmatrix} e^{i\theta} \\ e^{-i\theta} \end{pmatrix} \right|^2$$

$$= \tfrac{1}{2}|e^{i(\theta-\phi)} + e^{-i(\theta-\phi)}|^2$$

$$= \cos^2(\theta-\phi).$$

The physical prediction is the *same.*

There is an infinity of choice of representations all leading to the same physical predictions; here we have given two. In any ordinary application of quantum mechanics to a given physical situation, we must choose *one* of the possible representations and stick to it.

Arbitrary phase factors

Consider the states S, S′, represented respectively by the vectors v and $v' = v\,e^{i\theta}$, where θ is a number which will have to be *real* in order that both v and v' should be normalized. Rule II says that for a state with vector w

the probability that O(S) yields 'yes' is $|w^+v|^2$, while

the probability that O(S′) yields 'yes' is $|w^+v\,e^{i\theta}|^2$.

Whatever our choice of w, these two results are identical. Consequently, the states S and S′ are indistinguishable, and it would therefore be wrong to regard them as different states. We conclude that if S may be represented by the vector v, it may equally well be represented by the vector $v\exp(i\theta)$, with θ any real number; the physical predictions (which are what really matter) will be the same. The factor $\exp(i\theta)$ is an **arbitrary phase factor.**

We can now say how many complex components a state vector must have in any particular case. A vector with n complex components depends on $2n$ real parameters, of which $(2n-1)$ are independent (by the normalization requirement). One of these parameters may be used to specify the arbitrary phase factor, leaving $(2n-2)$ to specify the actual physical state.

PROBLEM

The *general* state of polarization (plane, circular, or elliptic) of a collimated monochromatic beam of light requires *two* real parameters to specify it. Deduce that the quantum description of photon polarization must use a vector with two complex components.

Circularly polarized light

To illustrate the rules, we shall search for the state vector representing a circularly polarized photon, assuming the representation for plane-polarization which we have used earlier (p. 8):

$$P_\theta \leftarrow \begin{pmatrix} \cos\theta \\ \sin\theta \end{pmatrix}.$$

Experiment shows that, on meeting a plane-polarizer in any orientation, a circularly polarized beam loses exactly half of its intensity; thus in quantum language a circularly polarized photon gets through with probability $\frac{1}{2}$.

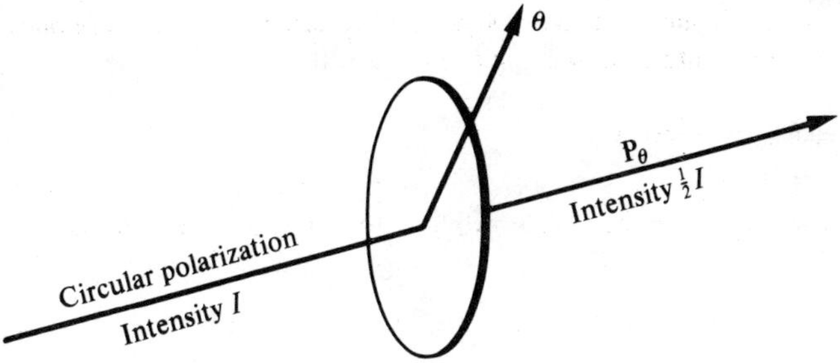

Alternatively, we may say that making the simple observation $O(P_\theta)$ will yield 'yes' with probability $\frac{1}{2}$.

Now suppose that the state of circular polarization is represented by the state vector $\begin{pmatrix} a \\ b \end{pmatrix}$ which is to be found. Then Rule II (p. 8) gives

$$\tfrac{1}{2} = \left|(\cos\theta, \sin\theta)\begin{pmatrix} a \\ b \end{pmatrix}\right|^2 \qquad \text{for all } \theta,$$

or

$$\tfrac{1}{2} = |a\cos\theta + b\sin\theta|^2 \qquad \text{for all } \theta.$$

This can be achieved only if $|a| = |b| = 1/\sqrt{2}$ and $b = \pm ia$ (problem 1, below). By a suitable choice of arbitrary phase factor we may take $a = 1/\sqrt{2}$, thus obtaining two vectors

$$\frac{1}{\sqrt{2}}\begin{pmatrix}1\\ \mathrm{i}\end{pmatrix} \quad \text{and} \quad \frac{1}{\sqrt{2}}\begin{pmatrix}1\\ -\mathrm{i}\end{pmatrix}.$$

These in fact give exactly the two states of circular polarization, one left and one right. (Which is which depends on conventions about left- and right-handedness not considered till later.)

PROBLEMS

1. Show that

$$2|a\cos\theta + b\sin\theta|^2 = |a|^2 + |b|^2 + (|a|^2 - |b|^2)\cos 2\theta + (a^*b + b^*a)\sin 2\theta,$$

and that the value of this expression is 1 for all θ if and only if $a = \mathrm{e}^{\mathrm{i}\alpha}/\sqrt{2}$, $b = \pm\mathrm{i}\,\mathrm{e}^{\mathrm{i}\alpha}/\sqrt{2}$, where α is real but otherwise arbitrary. (The coefficients of the oscillatory terms must be zero.)

2. For the alternative representation given on p. 10, find the state vectors which represent the two states of circular polarization. (These state vectors must each satisfy

$$\frac{1}{2} = \left|\frac{1}{\sqrt{2}}(\mathrm{e}^{-\mathrm{i}\theta}, \mathrm{e}^{\mathrm{i}\theta})\begin{pmatrix}c\\ d\end{pmatrix}\right|^2$$

for all θ.)

2. Observables

Why we need observables

MANY observations in physics are concerned with the *measurements of magnitudes*: the outcome of the measurement is a real number, rather than 'yes' or 'no'. A physical quantity which may be measured in this way is often called an *observable*. At first sight it may seem that our scheme of 'simple' observations (p. 8) is too restricted to allow for measurements of this kind. A measurement can be related to simpler observations, however. The question

'What is the magnitude of the physical quantity X?'

is rewritten as a set of simple questions

is ξ_1 the magnitude of X?
is ξ_2 the magnitude of X?
... and so on,

where the real numbers $\xi_1, \xi_2, \ldots$ are the possible outcomes of the measurement of X. The answer to each of these questions is 'yes' or 'no'. It would be tedious to decompose every question in this way, and we need to develop the notation. It turns out that the natural way to represent physical entities like X is by a *matrix*.

Mathematical prerequisites

An $n \times n$ **matrix** A is a set of n^2 **elements** (real or complex numbers),

$$A_{ij} \qquad (i = 1, 2, \ldots, n; j = 1, 2, \ldots, n).$$

(Sometimes A is conventionally presented as a square array, with rows and columns indexed by i and j respectively,

$$A = \begin{pmatrix} A_{11} & A_{12} & \cdots & A_{1n} \\ A_{21} & A_{22} & \cdots & A_{2n} \\ \cdots & \cdots & \cdots & \cdots \\ A_{n1} & A_{n2} & \cdots & A_{nn} \end{pmatrix}.$$

This is not often done in quantum mechanics, unless n is small.) The **Hermitian conjugate** of A, denoted by A^+ ('A-dagger') is the *complex conjugate of the transpose of* A; as an array

$$A^+ = \begin{pmatrix} A_{11}^* & A_{21}^* & \cdots & A_{n1}^* \\ A_{12}^* & A_{22}^* & \cdots & A_{n2}^* \\ \cdots & \cdots & \cdots & \cdots \\ A_{1n}^* & A_{2n}^* & \cdots & A_{nn}^* \end{pmatrix}.$$

Thus

$$(A^+)_{ij} = A_{ji}^*.$$

A **Hermitian** matrix is its own Hermitian conjugate.

The **zero matrix** has all elements equal to zero, and is usually written 0. The fact that the same symbol is used for zero, a zero vector, and a zero matrix does not cause confusion in practice; the appropriate interpretation is always clear from the context.

The **unit matrix** has elements

$$\delta_{ij} = \begin{cases} 1 & (i = j), \\ 0 & (i \neq j) \end{cases}$$

(the 'Kronecker delta'); as an array it is

$$\begin{pmatrix} 1 & 0 & \dots & 0 \\ 0 & 1 & \dots & 0 \\ \dots & \dots & \dots & \dots \\ 0 & 0 & \dots & 1 \end{pmatrix}.$$

It is usually written 1; again any possible ambiguity is resolved by the context.

The **sum** of two $n \times n$ matrices A and B is obtained by adding corresponding elements:

$$(A+B)_{ij} = A_{ij} + B_{ij}.$$

We may multiply a matrix by a complex number λ; each element must be individually multiplied by λ,

$$(\lambda A)_{ij} = \lambda \, . \, A_{ij}.$$

Suppose that A, B are $n \times n$ matrices, v is an n-component column vector, and w^+ is an n-component row vector. There are three important versions of the matrix **product**: the product of two matrices has elements defined by

$$(AB)_{ij} = \sum_{k=1}^{n} A_{ik} B_{kj} \quad \text{(itself a matrix)};$$

the product of a matrix and a column vector

$$(Av)_i = \sum_{k=1}^{n} A_{ik} v_k \quad \text{(itself a column vector)};$$

the product of a row vector and a matrix

$$(w^+ A)_i = \sum_{k=1}^{n} w_k^+ A_{ki} \quad \text{(itself a row vector)}.$$

These are different manifestations of the general rule of matrix multiplication: the elements of the product are (when written as an array) the scalar products of the rows of the first factor with the columns of the second.

For any $n \times n$ matrix A we may write down an equation of central importance in quantum mechanics: if v is a vector and λ a number (possibly complex)

$$Av = \lambda v, \qquad \text{i.e.} \sum_{j=1}^{n} A_{ij} v_j = \lambda v_i \quad (i = 1, \dots, n).$$

For given A, λ this is a system of n linear homogeneous equations in the n unknowns $v_1, v_2, \dots, v_n$. Generally, the only solution of these equations is $v_i = 0$, all i; for certain **eigenvalues** λ, however, a non-zero solution v exists (the **column eigenvector** corresponding to the eigenvalue). The determinant of the system of equations is a polynomial $\Delta(\lambda)$ of degree n, and the eigenvalues are the zeros of this polynomial. Thus there is at least one eigenvalue and corresponding eigenvector, and there may be as many as n (*eigen* is German for 'own'; the eigenvalues are *special* to the matrix).

We may also write down an equation for **row eigenvectors,**

$$w^+ A = \mu w^+.$$

The possible values for μ are the same as those for λ, but in general there is no simple relation between the corresponding eigenvectors w^+ and v.

There is an **orthogonality relation** between the row and column eigenvectors.

$$w^+ v = 0 \qquad \text{whenever } \mu \neq \lambda.$$

When A is *Hermitian*, there are certain important additional results. First, all its eigenvalues are *real*. Also, if v is a column eigenvector, then its Hermitian conjugate v^+ is a row eigenvector with the same eigenvalue. If v_1, v_2 are two eigenvectors with eigenvalues λ_1, λ_2, then we have the important orthogonality relation

$$v_1^+ v_2 = 0 \qquad \text{whenever } \lambda_1 \neq \lambda_2.$$

A **complete orthonormal set** of n-component vectors is a set $v_1, v_2, \ldots, v_n$, satisfying the 'orthonormality relations'

$$v_i^+ v_j = \delta_{ij} \equiv \begin{cases} 1 & \text{if } i = j, \\ 0 & \text{if } i \neq j, \end{cases}$$

$i = 1, 2, \ldots, n$; $j = 1, 2, \ldots, n$. Any other n-component vector u whatever may be expressed in exactly one way as a linear combination (**superposition**) of the vectors of a given complete orthonormal set

$$u = \sum_{i=1}^{n} c_i v_i.$$

The coefficients in the superposition are given by $c_i = v_i^+ u$.

A Hermitian $n \times n$ matrix always has as many as n linearly independent eigenvectors. It is always possible to select a complete orthonormal set of eigenvectors $v_1, v_2, \ldots, v_n$; this choice is never unique, since each v_i may be multiplied by an arbitrary phase factor without altering the orthonormality. If also the corresponding eigenvalues $\lambda_1, \lambda_2, \ldots, \lambda_n$ are not all distinct, there is an even greater arbitrariness in the choice, on account of **degeneracy**.

If we have at our disposal any complete orthonormal set of eigenvectors $v_1, v_2, \ldots, v_n$ of the Hermitian matrix A, with eigenvalues $\lambda_1, \lambda_2, \ldots, \lambda_n$, we may write down an **eigenvalue–eigenvector decomposition** of A:

$$A = \sum_{i=1}^{n} v_i \lambda_i v_i^+.$$

This expression is to be interpreted as follows: if u is a column vector then the product

$$Au = \left\{ \sum_{i=1}^{n} v_i \lambda_i v_i^+ \right\} u$$

$$= \sum_{i=1}^{n} v_i (\lambda_i v_i^+ u).$$

This is a superposition of the vectors v_i with coefficients $\lambda_i v_i^+ u$, which are, of course, ordinary numbers.

If the function $f(z)$ is defined for $z =$ each λ_i, the **matrix function** $f(A)$ is defined by

$$f(A) = \sum_{i=1}^{n} v_i f(\lambda_i) v_i^+;$$

when $f(z)$ is a polynomial, or power series convergent for each λ_i, this agrees with the result obtained by applying the rules for matrix addition and multiplication.

There are two important special cases. First, if $f(z) = 1$, we obtain

$$\sum_{i=1}^{n} v_i v_i^+ = 1 \qquad \text{(unit matrix).}$$

Second, if $f(\lambda_i) = 0$ for each λ_i, then $f(A) = 0$; in particular, A satisfies its own characteristic equation $\Delta(A) = 0$.

The possible outcomes of a measurement

We are now ready to incorporate *measurements* into quantum mechanics. There are different ways of approaching this kind of problem, and we shall follow one which uses the idea of a *mean value* of a physical observable. This provides a fairly direct route to our goal, and is particularly useful when we come to consider the *correspondence principle*, which relates certain results in quantum mechanics with analogous results in classical mechanics.

First of all, it will be useful to establish certain facts about the possible outcomes of an experimental act of measurement. It should be emphasized that we still have in mind systems whose states may be represented by state vectors with n components, n being finite. Some of our conclusions will need to be modified later.

There is a limit to the number of distinct possible outcomes of a measurement. Consider a quantity X for which each of the questions 'is ξ_j the magnitude of X?' is a '*simple*' observation $O(v_j)$ (p. 8). We shall now show that the states $v_1, v_2, \ldots$ form an orthonormal set. Suppose the system is in one of these special states, v_1, say. Then the answer to

'is ξ_1 the magnitude of X?'

is certainly 'yes', while the answer to

'is ξ_k the magnitude of X?'

is in consequence certainly 'no', provided $k \neq 1$. Alternatively

$$\left.\begin{array}{r}\text{the probability that}\\ \text{measuring } X \text{ yields } \xi_k\end{array}\right\} \text{ is } \begin{cases} 1 & \text{if } k = 1, \\ 0 & \text{if } k \neq 1. \end{cases}$$

Now, by Rule II (p. 8), we infer that

$$|v_1^+ v_k| = \begin{cases} 1 & \text{if } k = 1, \\ 0 & \text{if } k \neq 1. \end{cases}$$

We could have started the argument with *any* of the special states instead of v_1, and we finally conclude that

$$|v_j^+ v_k| = \begin{cases} 1 & \text{if } j = k, \\ 0 & \text{if } j \neq k \end{cases}$$

for *any* pair of the special state vectors v_j, v_k. That is, the state vectors form an orthonormal set.

Now an orthonormal set of n-component vectors cannot have more than n members. It follows that there cannot be more than n different possible outcomes of a measurement of X.

This conclusion is well confirmed by experiment. For example, when a pencil beam of unpolarized light falls on a birefringent material (e.g. a crystal of calcite), it becomes *two* beams inside the material, and these beams are

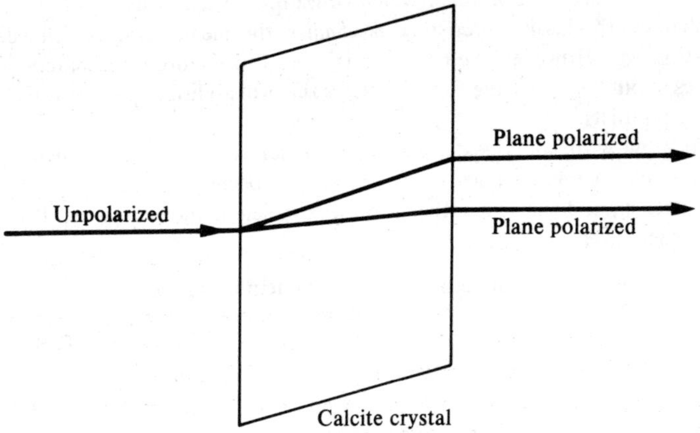

plane-polarized at right-angles to each other. It is *as if* the crystal 'asks' the question 'in which of two directions is the incoming photon polarized?' and then sends the photon into one beam or the other according to the answer. We know of no material which will produce more than two beams in similar circumstances, and for this reason we are led to assume that a two-component vector will describe photon polarization adequately.

PROBLEM

A pencil beam of light falls on a slab of birefringent material, and emerges as two parallel beams. These beams fall on a second differently oriented slab of the same material; each separates into two, and therefore *four* parallel beams emerge from the second slab. Explain why this does not contradict the foregoing discussion.

Mean values

Imagine that a system is in the state u and that a measurement of the physical quantity X is made, with possible outcomes $\xi_1, \xi_2, \ldots$; also that this experiment is repeated for the same state u a very large number of times and that the result ξ_j occurs a proportion p_j of the times. Then we may define a **mean value** of X for the state u by

$$\langle X \rangle_u = \sum_j p_j \xi_j.$$

(The mean value is sometimes called the **average value, expectation value**, or just **expectation.** Note that it need not be equal to any of the possible values ξ_j.)

So far we have said nothing which could not be said also about *classical* mechanics. In classical *statistical mechanics*, the mean value is defined in just this way. However, the next step is special to quantum mechanics, and comes about because Rule II (p. 8) relates the probabilities p_j to the values of scalar products.

To begin with, we consider a slightly special case. If X has the *maximum* number n of possible values, then each value ξ_i occurs with certainty for just one state v_i, and the v_i together form a complete orthonormal set. Rule II (p. 8) then gives

$$\begin{aligned} p_j = \text{probability of getting } \xi_j \text{ for the system in state } u &= |v_j^+ u|^2 \\ &= u^+ v_j v_j^+ u. \end{aligned}$$

Thus

$$\begin{aligned} \langle X \rangle_u &= \sum_j u^+ v_j \xi_j v_j^+ u \\ &= u^+ X u \end{aligned}$$

where (in accord with custom) we have used the *same* symbol X to denote the Hermitian matrix whose eigenvalue–eigenvector decomposition is

$$X = \sum_j v_j \xi_j v_j^+.$$

Also in accord with custom, we refer to both the physical quantity X and the matrix X which **represents** it as the **observable** X.

The observable X has the important properties

(1) its eigenvalues (p. 14) are the possible values of the physical quantity X; and

(2) its eigenvectors represent those **eigenstates** for which the corresponding eigenvalues are *certain* to be observed.

Even when an observable does not have the maximum number n of possible values, these properties are still true, and may if desired be taken as the defining properties of an observable.

PROBLEMS

1. Show that the number of possible values of a physical observable X is the largest possible if and only if the matrix which represents X is non-degenerate.

2. A 'simple' observation can be made into an observable by assigning numerical values 'yes' ≡ 1 and 'no' ≡ 0. Show that when this is done the matrix (using the representation on p. 8)

$$\begin{pmatrix} \cos^2\theta & \cos\theta\sin\theta \\ \cos\theta\sin\theta & \sin^2\theta \end{pmatrix}$$

represents a 'simple' observation, and show that its mean value for the state P_ϕ is $\cos^2(\theta-\phi)$. What is the relevance of this to experimental results? Describe a suitable apparatus for making this observation. (The eigenvectors of the matrix must be the state vectors for P_θ and $P_{\theta+\pi/2}$, with respective eigenvalues 1,0.)

3. Using the alternative representation (p. 10) show that the matrix

$$\tfrac{1}{2}\begin{pmatrix} 1 & e^{2i\theta} \\ e^{-2i\theta} & 1 \end{pmatrix}$$

represents a 'simple' observation, and show that its mean value for the state P_ϕ is $\cos^2(\theta-\phi)$. Relate this result to that of problem 2.

4. A **projection operator** is a Hermitian matrix whose eigenvalues are 0 or 1. Show that with the assignment 'yes' = 1 and 'no' = 0 every 'simple' observation is represented by a projection operator, but that not every projection operator represents a 'simple' observation. (A projection operator whose eigenvalue 1 is degenerate does not represent a 'simple' observation.)

The correspondence principle

Up till now, mean values have simply provided a convenient starting point for our discussion of observables, and it is generally true that the mean value of an observable chosen at random is of not much physical significance.

However, there are a few physical entities which are *naturally additive.* For example, the total linear momentum of a set of particles is just the vector sum of the individual linear momenta of the particles, and the same is true of their total energy provided there is no mutual interaction. The vector position of the centre of mass of a set of particles is the *mean* of their individual vector positions.

There are not many such entities, but they provide an important link between quantum and Newtonian mechanics. Without such a link, it would not be possible to identify in quantum mechanics the entities which correspond correctly to the fundamentally important concepts of classical mechanics, such as energy, momentum, and the like. The necessary link may be established at many different levels of generality, some requiring a considerable familiarity with rather advanced topics in classical mechanics. In this book we shall be content with basically simple applications of this *correspondence principle* to special cases. In practice, applying the principle always involves looking at some macroscopic system or other from *two* points of view, first as an 'everyday' classical system, and then as an assemblage of a large number of quantum systems. Both viewpoints must then give the same result, and this fact provides valuable information.

To see how this comes about, consider some large object (a billiard ball, say), made up of an almost unimaginably large number of very small atomic particles, each subject to *quantum* laws. The large-scale behaviour of such an object is governed by Newton's laws in a well-known way; e.g. the centre-of-mass motion is governed by an equation of motion relating total momentum and total force—both naturally additive quantities. Quantum mechanics, as applied to all the constituent particles, must predict this Newtonian equation correctly; this requirement of consistency is known as the **correspondence principle,** and was first put forward for a special case by Bohr in 1913, in connection with the spectrum of atomic hydrogen. Bohr asserted that if the electron is in an orbit of large radius then it will *radiate* according to the classical rules for a moving charge.

The angular momentum of a photon

To illustrate the correspondence principle idea in action we shall consider the angular momentum carried by a photon. To begin with, we think classically.

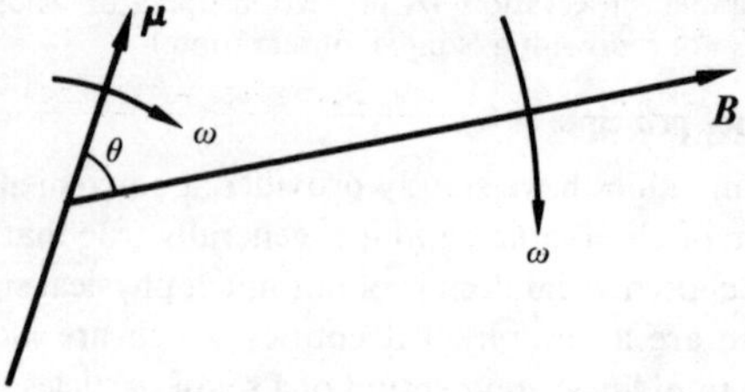

Imagine a beam of right-handed circularly polarized light of angular frequency $\omega(=2\pi\nu)$ falling perpendicularly on to the figure; the magnetic vector $\boldsymbol{B}$ in the plane of the figure will then rotate clockwise with angular velocity ω. Also imagine a small magnetic dipole $\boldsymbol{\mu}$ rotating with angular velocity ω, but lagging the vector $\boldsymbol{B}$ by the constant angle θ. This is a familiar situation of a kind of idealized synchronous motor.

The rate at which angular momentum is transferred to the dipole is (by Newton's laws) the torque G acting on the dipole (in fact, $G = \mu B \sin\theta$). Also, the rate at which this torque does work on the dipole is $G\omega$. Hence we have

$$\frac{\text{rate of energy transfer to the dipole}}{\text{rate of angular momentum transfer to the dipole}} = \omega.$$

By the correspondence principle, this rather general result must be predicted by quantum mechanics also.

Now we shall adopt the quantum point of view and regard the light beam as a beam of photons; we suppose that N photons are absorbed in unit time. We know that each photon carries an energy $h\omega/2\pi$, and that energy is naturally additive; in consequence,

$$\text{rate of energy transfer to the dipole} = Nh\omega/2\pi.$$

Consistency with the classical picture now requires that

$$\text{rate of angular momentum transfer to the dipole} = Nh/2\pi.$$

Since angular momentum also is naturally additive, we conclude

> the mean angular momentum carried by one right-handed circularly polarized photon is $\hbar \equiv h/2\pi$.

(The unit $\hbar \equiv h/2\pi$ is a natural unit for all quantum-mechanical angular momenta, and is of fundamental importance. Therefore it is given a notation of its own.) A more complete discussion (problem 3, p. 22) may be used to show that the angular momentum carried by one right-handed circularly polarized photon is *precisely* $\hbar$ (or $-\hbar$ for a left-handed circularly polarized photon).

We may use this conclusion to obtain the observable for photon angular momentum in any representation of our choice. In the representation for which $P_\theta \leftarrow (\cos\theta, \sin\theta)$, we must draw up the scheme (cf. p. 12)

state	eigenvalue	corresponding normalized eigenvector
right-handed circular	$\hbar$	$\frac{1}{\sqrt{2}}\begin{pmatrix}1\\ i\end{pmatrix}$
left-handed circular	$-\hbar$	$\frac{1}{\sqrt{2}}\begin{pmatrix}1\\ -i\end{pmatrix}$

which leads to the observable (p. 18, bottom)

$$\textbf{angular momentum} \text{ of a photon} = \hbar\begin{pmatrix}0 & -i\\ i & 0\end{pmatrix}.$$

Note how the *classical* idea 'angular momentum' has been transferred in a natural way to a *quantum* observable with the help of the correspondence principle.

PROBLEMS

1. *Plane*-polarized light of angular frequency ω falls on a dipole rotating clockwise with angular velocity ω. Show that

$$\langle\text{energy transfer}\rangle = \omega\,\langle\text{angular momentum transfer}\rangle,$$

where $\langle\ldots\rangle$ here means *time* average.

2. Evaluate the mean angular momentum of a photon in a plane-polarized beam (zero). In problem 1, where then does the angular momentum transferred come from? What can you say about the polarization of the beam *after* it has encountered the rotating dipole?

3. A more complete (classical) discussion of the angular momentum properties of polarized light gives the following results. Suppose a beam of light falls on a perfectly black (i.e. perfectly absorbing) surface. The surface will be heated (on account of the energy it absorbs) and may experience a torque (on account of the angular momentum it absorbs). Then, if the beam is plane-polarized (P_θ) in any orientation, the torque is zero, while, if the beam is circularly polarized ($C_\pm$),

$$\frac{\text{rate of energy transfer}}{\text{torque}} = \omega.$$

Use the correspondence principle to infer that if M is the observable for the angular momentum of one photon then

$$\langle M \rangle_{P_\theta} = 0, \quad \text{all } \theta, \quad \text{and} \quad \langle M \rangle_{C_\pm} = \pm\hbar. \tag{2.1}$$

Hence show that, in the representation where

$$P_\theta \leftarrow \begin{pmatrix} \cos\theta \\ \sin\theta \end{pmatrix} \qquad C_\pm \leftarrow \frac{1}{\sqrt{2}}\begin{pmatrix} 1 \\ \pm i \end{pmatrix},$$

these requirements lead *uniquely* to

$$M = \hbar\begin{pmatrix} 0 & -i \\ i & 0 \end{pmatrix}.$$

(Take for M the general Hermitian matrix $\begin{pmatrix} a & c \\ c^* & b \end{pmatrix}$, with a and b real, and examine the implications of the requirements (2.1) for the elements of this matrix.)

Uncertainty

The fundamental contrast between the classical and the quantum approach is well illustrated by the work of the last section. Suppose a photon is in a definite state, with mean angular momentum a. Then the two points of view are:

Classical: a system in a definite state will yield a predictable result for any observation made on it. If the photon has mean angular momentum a, this can come about only if the photon has angular momentum of precisely a.

Quantum: measuring the angular momentum of the photon will yield either $+\hbar$ or $-\hbar$. Even though the photon is in a definite state, it does *not* follow that any observation will yield a predictable result. The results $\hbar$, $-\hbar$ will be observed for proportions p, $(1-p)$ of all occasions, with p such that the mean $= p\hbar-(1-p)\hbar = a$.

Though the *mean* results are the same, the actual observations may clearly be very different.

In quantum mechanics the results of an observation repeated many times on the same state may show a *scatter* about their mean value; this scatter is quantified as follows. First we note that if u is an eigenstate of an observable X, then the corresponding eigenvalue is $\langle X\rangle_u$ (problem 4, below); i.e.

$$(X - \langle X\rangle_u)u = 0.$$

In fact, this equation is satisfied if and only if u is a normalized eigenstate of X, i.e. if and only if measuring X in the state u gives a particular result with certainty. The extent to which this equation fails to be satisfied for a *general* state u therefore gives a measure of the unpredictability of the result of measuring X. With this in mind we define the **uncertainty** $\Delta_{X,u}$ by

$$\begin{aligned}\Delta^2_{X,u} &= u^+(X-\langle X\rangle_u)(X-\langle X\rangle_u)u, \qquad \Delta_{X,u} \geqslant 0,\\ &= \langle (X-\langle X\rangle_u)^2\rangle_u\end{aligned}$$

or, on rearrangement,

$$\Delta_{X,u} = +\sqrt{(\langle X^2\rangle_u - \langle X\rangle_u^2)}.$$

This happens to be the *standard deviation* (in the usual statistical sense) of the observed values of X from their mean value when the system is in the state u. In classical mechanics, uncertainty is an irrelevance (being always zero). In quantum mechanics, $\Delta_{X,u}$ is zero if and only if u is an eigenvector of X, and is a maximum for any state where the two extreme values of X are equally likely. It therefore corresponds reasonably closely to an intuitive idea of uncertainty.

PROBLEMS

1. Justify the rearrangement

$$u^+(X-\langle X\rangle_u)(X-\langle X\rangle_u)u = \langle X^2\rangle_u - \langle X\rangle_u^2.$$

2. A photon in state P_θ falls on a polarizer set at angle ϕ. It either survives (result $= 1$) or is absorbed (result $= 0$). Show that the uncertainty in the result is

$$\tfrac{1}{2}|\sin 2(\theta-\phi)|.$$

$$\left(\text{In this case } X = \begin{pmatrix}\cos^2\theta & \cos\theta\sin\theta\\ \cos\theta\sin\theta & \sin^2\theta\end{pmatrix};\ \text{see problem 2, p. 19.}\right)$$

3. Show that the maximum possible uncertainty for any observable X is $\frac{1}{2}|x_1-x_2|$, where x_1 and x_2 are the extreme eigenvalues of X. (Maximize $\sum_i p_i x_i^2 - (\sum_i p_i x_i)^2$ with respect to p_i subject to $\sum_i p_i = 1$ and each $p_i \geqslant 0$.)

4. If ξ, u are eigenvalue and normalized eigenvector of X, show that $\xi = \langle X\rangle_u$. (Note that $u^+(Xu-\xi u) = 0$.)

Uncertainty relations

There exists an important inequality for uncertainties of observables. Suppose A and B are two observables and that all mean values are taken for a fixed state vector u. The expression

$$\langle (A+\mathrm{i}\lambda B)(A-\mathrm{i}\lambda B)\rangle$$

is the squared length of the vector $(A-\mathrm{i}\lambda B)u$ and is therefore always real and non-negative for any real λ; when rearranged, it is

$$\lambda^2\langle B^2\rangle - \mathrm{i}\lambda\langle (AB-BA)\rangle + \langle A^2\rangle,$$

or

$$\lambda^2\langle B^2\rangle - \mathrm{i}\lambda\langle [A, B]\rangle + \langle A^2\rangle,$$

where we have written $[A, B]$ for the **commutator** $AB-BA$; note that $\mathrm{i}[A, B]$ is Hermitian and that $\langle \mathrm{i}[A, B]\rangle$ is real (problem 1, below). Here we have a quadratic expression which is negative for no value of λ; the condition for this is '$b^2-4ac \leqslant 0$', i.e.

$$\langle A^2\rangle\langle B^2\rangle \geqslant \tfrac{1}{4}|\langle [A, B]\rangle|^2.$$

In this general result set $A = X-\langle X\rangle$, $B = Y-\langle Y\rangle$, and note that since $\langle X\rangle$ and $\langle Y\rangle$ are numbers, $[A, B] = [X, Y]$ (problem 2, below). Then

$$\Delta_X\Delta_Y \geqslant \tfrac{1}{2}|\langle [X, Y]\rangle|.$$

This is an **uncertainty relation**; it imposes an important restriction on the uncertainties of two observables for the same state. Generally, if $[X, Y]$ is not zero ('if X and Y do not commute'), Δ_X and Δ_Y cannot *both* be made arbitrarily small simultaneously.

PROBLEMS

1. Show that $\mathrm{i}[A, B]$ is Hermitian, if A and B are.

2. Show that if x and y are numbers (i.e. if they are both multiples of the unit matrix) then

$$[X-x, Y-y] = [X, Y].$$

3. An alternative route to the uncertainty relation is suggested as follows. Write

$$\phi(\lambda) = \lambda^2\langle B^2\rangle - \mathrm{i}\lambda\langle [A, B]\rangle + \langle A^2\rangle,$$

and find that value of λ for which $\phi(\lambda)$ takes its minimum value (the condition is $\partial\phi(\lambda)/\partial\lambda = 0$). Substitute this value of λ into $\phi(\lambda)$ to give the minimum value:

$$(\langle A^2\rangle\langle B^2\rangle - \tfrac{1}{4}|\langle A, B\rangle|^2)/\langle B^2\rangle.$$

Finally, use the inequality $\phi(\lambda) \geqslant 0$ to obtain the uncertainty relation.

4. Similar methods may be used to obtain other inequalities. In the following, the starting point is to set $\partial\chi/\partial\alpha_1 = \partial\chi/\partial\alpha_2 = 0$, where the complex $\alpha = \alpha_1 + \mathrm{i}\alpha_2$. Note however that a 'slick' way of doing this is to require $\partial\chi/\partial\alpha = \partial\chi/\partial\alpha^* = 0$, where the differentiation with respect to α (or α^*) is carried out as if α^* (or α) were kept constant.

By considering the minimum value of $\chi = \langle (A+\alpha^* B)(A+\alpha B)\rangle$ as α ranges over all *complex* values, show that $\langle A^2\rangle\langle B^2\rangle \geqslant |\langle AB\rangle|^2$. Hence show that

$$\Delta_X \Delta_Y \geqslant |\langle XY\rangle - \langle X\rangle\langle Y\rangle|.$$

What happens to this inequality when $X = Y$? (Since the range of α is wider than the range of λ was, this inequality may be expected to be more restrictive than the previous one. However, the one involving the commutator is frequently more useful, and we shall meet it again later.)

Incompatibility

There is yet another feature of quantum mechanics, closely connected with uncertainty, which has no analogue in classical mechanics. It may happen that two different acts of observation may be **incompatible** in the sense that they cannot be performed simultaneously. Here we are not talking about mere difficulty or inconvenience, but about a fundamental impossibility in principle.

When a measurement of an observable X is made, the system then goes into an eigenstate of X (p. 18). Similarly, when a measurement of Y is made, the system goes into an eigenstate of Y. It follows that simultaneous measurements of X and Y will send the system into a state which is simultaneously an eigenstate of both X and Y. Such a simultaneous eigenstate need not exist; in that case, the simultaneous measurement of X and Y is in principle not possible. This is the incompatibility referred to above.

We have already seen (p. 18) the important role played by the complete orthogonal set of eigenvectors of X in the theory of the measurement of X. The relevance of this to the problem of incompatibility lies in the following assertion: the observables X and Y may be measured simultaneously *if and only if they have a complete orthogonal set of eigenvectors in common.* Moreover, X and Y have a complete orthogonal set of common eigenvectors if and only if they **commute**, i.e.

$$XY = YX$$

or

$$[X, Y] \equiv XY - YX = 0$$

(problems 1 and 2, p. 26). We see that the commutator is important both for the uncertainty relation and for incompatibility; thus there is a close connection between the two.

A more informal approach to the topic of incompatibility is provided by considering what happens when a photon encounters a *sequence* of differently oriented plane-polarizers. For example, consider a photon in plane-polarization state P_0 which encounters a polarizer set at $\frac{1}{4}\pi$ and then a second polarizer at $\frac{1}{2}\pi$. The probability of surviving the first polarizer is $\cos^2 \frac{1}{4}\pi = \frac{1}{2}$; the probability of then surviving the second is $\cos^2(\frac{1}{2}\pi - \frac{1}{4}\pi) = \frac{1}{2}$. The overall probability of survival is thus $\frac{1}{4}$.

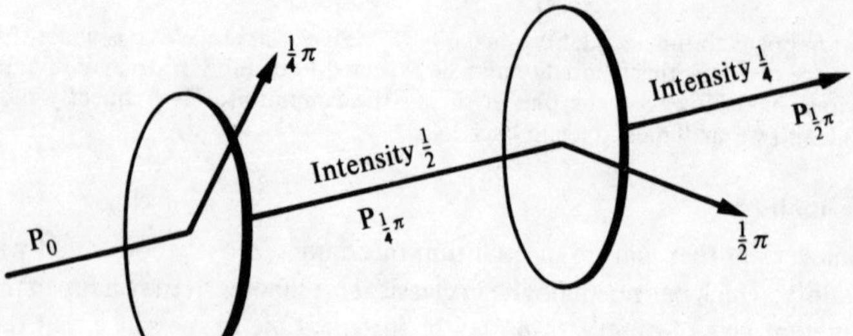

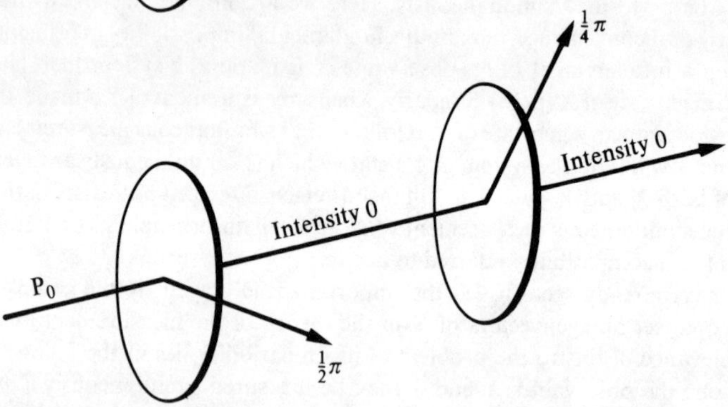

Now consider the situation, differing only in that the order of the two polarizers is reversed, the orientations remaining the same. In this case, the photon certainly does not survive even the first polarizer; the overall probability of survival is 0.

The two cases differ only in the order of the polarizers, and the predictions are different. Since the result depends on the order, it makes no sense, in principle, to ask for the result of passing the photon through both polarizers simultaneously. The point is not merely that the polarizers get in each other's way, but rather that *no* technique of observation can be devised which simultaneously measures plane-polarizations at $\frac{1}{4}\pi$ and $\frac{1}{2}\pi$. Of course, there is nothing particularly special about these angles; almost any angles will illustrate the point.

PROBLEMS

1. Show that if X and Y possess a complete orthogonal set of common eigenvectors, then they commute. (Show first that for any common eigenvector v, $(XY - YX)v = 0$. Deduce that $(XY - YX)u = 0$ for *any* vector and hence that $XY - YX = 0$.)

2. Show that, if X has no degenerate eigenvalues and X and Y commute, then each eigenvector of X is an eigenvector of Y. (If v_i are the eigenvectors of X, evaluate the left side of $v_j^+(XY-YX)v_i = 0$ to show that $v_j^+ Yv_i = 0$ if $i \neq j$. Then use $I = \sum_j v_j v_j^+$ to show that

$$Yv_i = IYv_i = \sum_j v_j(v_j^+ Yv_i) = v_i \times \text{constant.})$$

3. In an earlier problem (p. 19) an observable

$$\mathrm{P}(\theta) = \begin{pmatrix} \cos^2\theta & \cos\theta\sin\theta \\ \cos\theta\sin\theta & \sin^2\theta \end{pmatrix},$$

was introduced in connection with the plane-polarization of a photon. Evaluate $[\mathrm{P}(\theta), \mathrm{P}(\phi)]$, and show that it is zero only if $(\theta-\phi)$ is an integer multiple of $\frac{1}{2}\pi$. Interpret this result physically.

3. Equations of motion

Why we need equations of motion

THE major interest in experimental physics is to understand how states develop in time, or how they change under external influences. A typical experimental scheme is the following:

Preparation: A system is procured in a chosen state S_1;

Interaction: The system is allowed to develop, or is subjected to external 'forces';

Observation: An observation is made on the system in its now possibly different state S_2.

For example, in an atomic scattering experiment, 'preparation' may be obtaining a particle with definite linear momentum from an accelerator, 'interaction' may be collision with a target, and 'observation' may be the use of a suitable detector to measure the angle of deflection. Equations of motion provide the theoretical description of the interaction stage.

So far, these remarks are completely general, and may just as well be made for *classical* mechanics. It is when we ask for the *form* that equations of motion must take that the characteristic quantum features appear. The physical state of a quantum system is represented by a state vector; if this physical state varies in time, so does the state vector. It would be sufficient for our purposes to have a rule for obtaining the first time derivative of the state vector when the state vector itself is given; solving the resulting differential equation would then give the solution to the original physical problem.

Quantum dynamics

Consider a system whose column state vector $u(t)$ is not constant but varies with the passage of time t. If the variation is smooth enough (and in practical applications it always is), we may define the derivative with respect to time

$$\dot{u} \equiv \frac{\mathrm{d}u}{\mathrm{d}t}$$

in the simple sense that if

$$u = \begin{pmatrix} c_1(t) \\ c_2(t) \\ \vdots \end{pmatrix} \qquad \text{then } \dot{u} = \begin{pmatrix} \mathrm{d}c_1/\mathrm{d}t \\ \mathrm{d}c_2/\mathrm{d}t \\ \vdots \end{pmatrix}$$

(the state vector is differentiated component by component). An equation of motion for u is some rule for giving $\dot{u}$ when u is known; to find how u varies in time is then a matter of solving the resulting differential equations.

Out of a wide range of possibilities, it has been found in every case in quantum mechanics that the equation of motion takes the form

$$i\hbar\dot{u} = Hu \qquad \text{(Rule III)}$$

where H is a Hermitian matrix (see problem, below), depending on the properties of the system and of any external influence (and possibly even on the time t), but *always independent of* u; H is the **Hamiltonian** of the system, and provides a complete description of the system in the sense that nothing else is needed to determine the development of the system in time.

The Hamiltonian is a Hermitian matrix and is therefore an observable. The inclusion of the fundamental constant $\hbar$ in the equation implies that H has the dimensions of energy, and indeed it represents the energy of the system. (This statement will be justified in Chapter 4 with the help of the correspondence principle.)

W. Heisenberg (1925) and E. Schrödinger (1926) independently developed the first examples of quantum equations of motion in Hamiltonian form, for the quantum simple harmonic oscillator and the hydrogen atom respectively. Their approaches were apparently very different, but were soon shown to be equivalent. Since that time, the Hamiltonian form of the quantum equation of motion has hardly ever been questioned.

The name *Hamiltonian* harks back to a nineteenth century mathematician and astronomer W. R. Hamilton, who contributed enormously to the general theory of classical mechanics. The most complete expression of the correspondence principle consists in drawing out the analogy between classical mechanics in Hamilton's formulation and quantum mechanics in the form we are now developing. Though this analogy is one of the favourite points of departure in texts on quantum mechanics, we consider that it may be somewhat too advanced for this book, since the *classical* ideas—for a change—may be unfamiliar!

PROBLEM

Show that if u is normalized at $t = 0$, then it is normalized for all time. (First show that $i\hbar\dot{u}^+ = -u^+H$, and then evaluate $\mathrm{d}(u^+u)/\mathrm{d}t$. The point is that u remains normalized because H is Hermitian, and therefore a non-Hermitian Hamiltonian will not do.)

Optical activity

We are not quite ready to give an example of a quantum equation of motion; however, we shall now describe a very similar situation involving an equation of 'motion' where the independent parameter is not t but a space coordinate x.

Certain substances (e.g. quartz or turpentine) are *optically active*; a polarized beam of light passing through such a substance has its plane of polarization progressively rotated. In quantum language, a photon in state P_ϕ on entering the substance will be in state $P_{\phi+\alpha x}$ on travelling a distance x; the *optical activity* α depends only on the substance concerned.

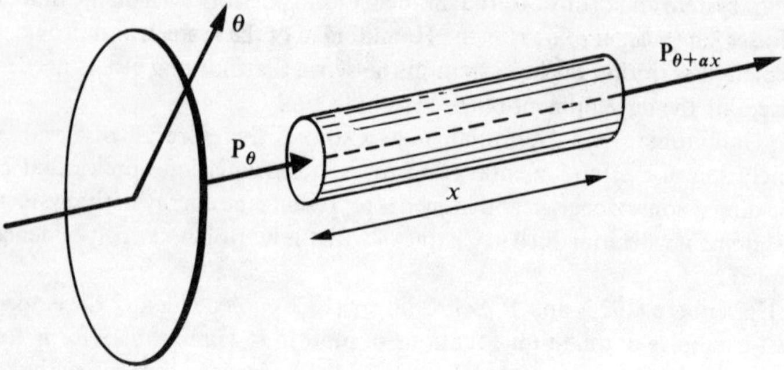

Using the familiar representation $P_\theta \leftarrow (\cos\theta, \sin\theta)$, we have that after travelling a distance x,

$$\text{the state of the photon is } P_{\phi+\alpha x} \leftarrow u(x) \equiv \begin{pmatrix} \cos(\phi+\alpha x) \\ \sin(\phi+\alpha x) \end{pmatrix}.$$

Differentiation with respect to x gives immediately

$$i\frac{du}{dx} = i\alpha\begin{pmatrix} -\sin(\phi+\alpha x) \\ \cos(\phi+\alpha x) \end{pmatrix} = Ku,$$

where K is the Hermitian matrix

$$K = \begin{pmatrix} 0 & -i\alpha \\ i\alpha & 0 \end{pmatrix},$$

describing *completely* the effect of the optical activity on the polarized photon.

We must emphasize the word 'completely'. We have obtained K by considering experimental results for *plane*-polarization only: we may now *use* K to *predict* the results for other polarizations. For example, let us ask what will happen to a *circularly* polarized photon whose state vector is initially

$$\frac{1}{\sqrt{2}}\begin{pmatrix} 1 \\ i \end{pmatrix} \equiv v(0).$$

Now this vector happens to be very simply related to K: it is an eigenvector of K with eigenvalue α. It is simple to verify (problem 2, p. 32) that as a result the solution of the equation of motion is

$$v(x) = \frac{1}{\sqrt{2}}\begin{pmatrix}1\\ i\end{pmatrix} e^{-i\alpha x}.$$

Clearly, $v(x)$ differs from $v(0)$ only by virtue of the phase factor $\exp(-i\alpha x)$, and thus the *physical* state of circular polarization is *not affected* by the optical activity. This is in complete accord with experiment. A similar result may be obtained for the state vector

$$\frac{1}{\sqrt{2}}\begin{pmatrix}1\\ -i\end{pmatrix},$$

which represents the other state of circular polarization.

Stationary states

There is nothing to prevent the Hamiltonian H itself depending on t. A quantum system subject to a non-constant external force, for example, will need a non-constant H for its description. However, the special case of a *time-independent* H frequently occurs, and is of fundamental importance.

Suppose H has an eigenvalue E and a corresponding eigenvector v. Then consider the system in the state v at time $t = 0$. At any other time t, the system must be in the state $u(t) = v\exp(-iEt/\hbar)$, as can be seen by direct substitution in the equation of motion (p. 29), using

$$i\hbar\dot{u} = Ev\exp(-iEt/\hbar) \quad \text{and} \quad Hu = Hv\exp(-iEt/\hbar);$$

these are equal since $Hv = Ev$.

Now the state vector $u(t)$ differs from v only by the phase factor $\exp(-iEt/\hbar)$; physically $u(t)$ is the same state for all time. Such a state is a **stationary state** (or **energy eigenstate**) of the system; any eigenstate of the Hamiltonian (assumed constant) is a stationary state. (Note that if the Hamiltonian varies in time, its eigenstates are not usually particularly important.)

The existence of stationary states in the theory has two very different practical consequences. Physically, such states have a permanence in time which aids their experimental investigation. Mathematically, they reduce the solution of the equation of motion to the eigenvalue problem for the Hamiltonian, and this is often a valuable simplification.

Note that the frequency of oscillation ν of the factor $\exp(-iEt/\hbar)$ is $\nu = E/2\pi\hbar = E/h$; thus $E = h\nu$. This is closely connected with the relation between the energy and the frequency of a photon and is the first suggestion that we are right to regard H as the observable for energy.

Since H is Hermitian, the eigenvectors v_j of H form a complete orthonormal set (p. 15); thus the vector representing a *general* state at time $t = 0$ may be expressed as a linear superposition of these eigenvectors

$$u(0) = \sum_j c_j v_j.$$

At time t we will have (as may be seen by substituting in the equation of motion)

$$u(t) = \sum_j c_j v_j \exp(-\mathrm{i}E_j t/\hbar);$$

i.e. the general solution is a linear combination of stationary solutions with constant coefficients c_j. In this sense, a knowledge of the stationary states provides the complete solution to the problem.

Note that the general solution contains oscillatory components with frequencies E_j/h, and no other frequencies. Allowing for the moment that the Hamiltonian is the observable for the energy of the system, the interpretation of the coefficients c_j is immediate:

$$\left.\begin{array}{l}\text{the probability that measuring the}\\ \text{energy of the system yields } E_j\end{array}\right\}\text{ is } |c_j|^2.$$

PROBLEMS

1. Show that if on entering the optically active medium the photon is in a general polarization state $\begin{pmatrix} c_1 \\ c_2 \end{pmatrix}$, with c_1 and c_2 possibly complex, then after travelling a distance x it will be in the state

$$\begin{pmatrix} c_1 \cos \alpha x - c_2 \sin \alpha x \\ c_1 \sin \alpha x + c_2 \cos \alpha x \end{pmatrix}.$$

2. Use the result of problem 1 to show that the two states of circular polarization are each physically unchanged on passing through an optically active medium.

3. If the Hamiltonian is constant in time, show that $u(t) = S(t)u(0)$ where $S(t)$ is a matrix function of H, namely, $S(t) = \exp(-\mathrm{i}Ht/\hbar)$. The analogous matrix for the optical activity example is

$$S(x) = \begin{pmatrix} \cos \alpha x & -\sin \alpha x \\ \sin \alpha x & \cos \alpha x \end{pmatrix};$$

show explicitly that $S(x) = \exp(-\mathrm{i}Kx)$. (See p. 15 for the definition of a matrix function.)

The equation of motion for a mean value

The mean value of an observable X for the state u may be expected to vary in time as u varies in time. In fact, we have (assuming that X is a *constant* Hermitian matrix)

$$\mathrm{i}\hbar \frac{\mathrm{d}\langle X \rangle_u}{\mathrm{d}t} = \mathrm{i}\hbar \frac{\mathrm{d}}{\mathrm{d}t}(u^+ X u) = \mathrm{i}\hbar(\dot{u}^+ X u + u^+ X \dot{u})$$

$$= -u^+ H X u + u^+ X H u$$

(here we have used the Hermitian conjugate equation $i\hbar\dot{u}^+ = -u^+H$, along with $H^+ = H$); hence

$$\frac{d}{dt}\langle X\rangle_u = \left\langle \frac{1}{i\hbar}[X, H]\right\rangle_u .$$

This expresses the rate of change of the mean value of X in terms of another mean value, that of the **commutator**

$$[X, H] \equiv XH - HX.$$

In the light of the correspondence principle, we expect equations of motion of this kind to provide a valuable link between the concepts of classical and quantum mechanics; there will be several examples in this book. The celebrated approach of Dirac was developed in 1925, and was based on the remarkable similarity of the commutator to the *Poisson bracket* of general classical mechanics; though extremely elegant, Dirac's methods are rather abstract and beyond the scope of this book.

When H is a constant matrix with eigenvalues E_j, the general state is represented by

$$u(t) = \sum_j c_j v_j \exp(-iE_j t/\hbar).$$

The mean value $\langle X\rangle_u$ is then

$$\begin{aligned}\langle X\rangle_u &= \sum_j c_j^* v_j^+ \exp(iE_j t/\hbar) \,.\, X \,.\, \sum_k c_k v_k \exp(-iE_k t/\hbar)\\ &= \sum_j \sum_k c_j^* c_k (v_j^+ X v_k) \exp\{i(E_j - E_k)t/\hbar\}.\end{aligned}$$

The time-variation of $\langle X\rangle_u$ is thus a superposition of simple oscillatory terms with frequencies $(E_j - E_k)/\hbar$, and no other. An application of this result to the quantum simple harmonic oscillator will be given later (p. 48).

PROBLEM

Show that if the Hamiltonian is a constant matrix, its mean value is constant. (This is the quantum version of *conservation of energy*.)

4. Quantum particles in one dimension: basics

What is a quantum particle?

SIMPLY, a quantum particle is a system which is as close as we can make it to a classical particle, i.e. a point-mass with the intuitive requirements

(1) it must be somewhere;
(2) it cannot be in two places at once;
(3) its motion must be continuous in time;
(4) it must obey Newton's laws of motion.

When we try to set up a quantum analogue it is easy to retain requirements (1) and (2). However, (3) must be drastically modified, while Newton's laws are now satisfied only by mean values of appropriate observables.

There are two reasons for studying quantum particles. The first is that they really do exist, at least in an approximation which ignores internal structure and certain relativistic effects: electrons, protons, mesons, etc. The second is on account of the correspondence principle which allows us in this particular case more than in any other to transfer important classical ideas like momentum, energy, force, and mass into the quantum realm.

Representing the state of a quantum particle

We shall consider the problem of a particle constrained to move on a straight line going to infinity in both directions. It is assumed that the particle must be somewhere, and cannot be in two places at once; alternatively put, if we make an *observation of the coordinate* x of the particle, we will certainly get a result. In line with the theory of observables of Chapter 2, we need an observable x whose eigenvalues are the possible results of the observation. These possible results will form a *continuous range* of values.

The immediate difficulty is that up to now all observables have been finite matrices with, therefore, a *finite* number of *distinct* eigenvalues. It is therefore time to generalize the formalism. Let us set up an *approximate* representation of the system by dividing the line into 'boxes' of length δ:

'box n' is the segment $n\delta \leqslant x < (n+1)\delta$ (n any integer).

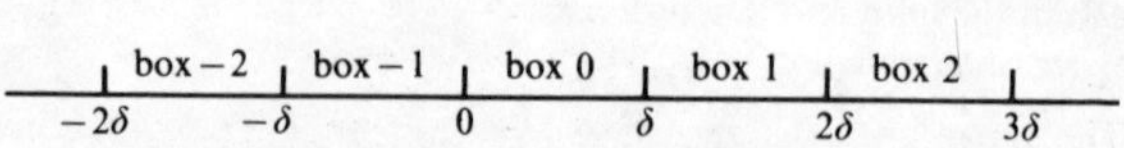

Then an approximate observation of x (to within a possible error δ) can be made by observing which box the particle is in. In line with the work of Chapter

1, we represent approximately the general state S of the system by a normalized column vector (now with an *infinite* number of complex components)

$$\mathrm{S} \leftarrow \begin{pmatrix} \vdots \\ c_{n-1} \\ c_n \\ c_{n+1} \\ \vdots \end{pmatrix}, \qquad \sum_{n=-\infty}^{\infty} |c_n|^2 = 1,$$

with the physical interpretation

$$\text{probability of finding the particle in 'box } n\text{'} = |c_n|^2.$$

To make the approximate representation exact, we must allow δ to tend to zero. Thus the boxes will become more numerous and shrink in size. As they shrink, the probability that the particle will be found in any particular box will tend to zero: it is roughly proportional to δ. Thus we expect that

$$\text{the } \textbf{probability density} \text{ at } x = \lim_{\delta \to 0} \delta^{-1} |c_{n(\delta)}|^2$$

should exist (here $n(\delta)$ is the box containing x; obviously it depends on δ). This suggests that we define the **wavefunction** for the state of the particle by

$$\psi(x) = \lim_{\delta \to 0} \delta^{-\frac{1}{2}} c_{n(\delta)}$$

and use this wavefunction as an appropriate way of representing S,

$$\mathrm{S} \leftarrow \psi(x).$$

Representing the state of a quantum particle by a wavefunction rather than by a column vector does not imply any change in basic principles. It is a natural development forced upon us merely by the fact that we wish to handle an observation whose possible outcomes form a continuous range.

PROBLEMS

1. Suppose that, in the limit $\delta \to 0$, the row vector $\delta^{-\frac{1}{2}} b_n^*$ and the column vector $\delta^{-\frac{1}{2}} c_n$ go over into the wavefunctions $\phi^*(x)$ and $\psi(x)$ respectively. Show that the scalar product $\sum_n b_n^* c_n$ goes over into

$$(\phi^*, \psi) \equiv \int_{-\infty}^{\infty} \phi^*(x) \psi(x) \, \mathrm{d}x.$$

2. Show that, in the limit $\delta \to 0$, a *normalized* vector c_n leads to a normalized wavefunction, i.e. a wavefunction ψ for which $(\psi^*, \psi) = 1$.

(Note: The above gives the correct generalization of the scalar product, and is important for the physical interpretation. If a 'simple' observation (p. 8) *certainly* yields the result ξ for the state ϕ and for no other state, then it will yield the result ξ for the state ψ with probability $|(\phi^*, \psi)|^2$, provided both ψ and ϕ are normalized.)

Schrödinger's equation for a free particle

A classical particle follows a continuous trajectory, but this cannot be said of a quantum particle. However, it is possible to retain continuity of motion in a modified sense by setting up the equation of motion properly.

Let us return to the approximate representation of the system (before $\delta \to 0$). The Hamiltonian for the system must be a matrix with an infinite number of rows and columns, one for each component c_n. If the typical element of H is H_{nm}, we may write the equation of motion for the state vector as a set of coupled differential equations

$$i\hbar \dot{c}_n = \sum_{m=-\infty}^{\infty} H_{nm} c_m, \qquad (n = \ldots, -2, -1, 0, 1, 2, \ldots).$$

We now introduce the idea of continuity in the form that what happens in box n depends directly only on what is happening in the *immediately adjacent* boxes. In particular, we say that $\dot{c}_n$ must depend only on c_{n-1}, c_n, c_{n+1} and on no other component:

$$i\hbar \dot{c}_n = H_{n,n-1} c_{n-1} + H_{nn} c_n + H_{n,n+1} c_{n+1}.$$

Each row of the Hamiltonian matrix now contains just three non-zero elements, the overwhelming majority of the elements being now zero.

To begin with, we shall assume that all boxes have the same properties. This will imply that the form of the equation for $\dot{c}_n$ will be independent of n, and, out of a number of physically equivalent possibilities, we shall set

$$H_{n,n-1} = H_{n,n+1} = -\hbar^2/2m\delta^2$$
$$H_{nn} = \hbar^2/m\delta^2 \qquad \text{(all } n\text{)}.$$

The parameter m is a 'property' of the quantum particle and has the dimensions of a mass; we shall anticipate some later developments by calling it the mass of the particle.

The δ^2 in the denominator of these expressions has been inserted with the limit $\delta \to 0$ in mind. We may now write the equation of motion in the form

$$i\hbar \frac{\mathrm{d}}{\mathrm{d}t}(c_n \delta^{-\frac{1}{2}}) = -\frac{\hbar^2}{2m} \cdot \frac{1}{\delta^2}(c_{n-1}\delta^{-\frac{1}{2}} - 2c_n\delta^{-\frac{1}{2}} + c_{n+1}\delta^{-\frac{1}{2}});$$

then with the approximations

$$c_n\delta^{-\frac{1}{2}} \sim \psi(x, t), \quad c_{n-1}\delta^{-\frac{1}{2}} \sim \psi(x-\delta, t), \quad c_{n+1}\delta^{-\frac{1}{2}} \sim \psi(x+\delta, t)$$

we take the limit $\delta \to 0$ to obtain

$$i\hbar \frac{\partial \psi}{\partial t}(x, t) = -\frac{\hbar^2}{2m} \frac{\partial^2 \psi(x, t)}{\partial x^2}.$$

This is **Schrödinger's equation** for the free particle with mass m in one dimension. It is the quantum equation of motion which describes how the state represented by ψ changes as time passes.

These developments show that matrix algebra is no longer enough for our purposes, and we need to introduce the more general idea of a linear operator. For example, in order to write the Schrödinger equation in the general form for a quantum equation of motion, namely,

$$i\hbar \frac{\partial \psi}{\partial t} = H\psi,$$

we must introduce as the Hamiltonian the *differential operator*

$$H = -\frac{\hbar^2}{2m}\frac{\partial^2}{\partial x^2}.$$

We shall see that many of the ideas of matrix algebra may be carried over to linear operators. In particular, we shall see later that H is an observable and represents the energy of the particle.

A particle in an external potential $V(x)$

In the last section, $H_{n,n-1}$, H_{nn}, and $H_{n,n+1}$ were all assumed to be independent of n; relaxing this leads to a physical system of a particle subject to an external influence changing according to its position. There are several possibilities, and one which is far more important than any other is obtained by setting

$$H_{n,n-1} = H_{n,n+1} = -\hbar^2/2m\delta^2, \qquad H_{nn} = \hbar^2/m\delta^2 + V_n \quad (\text{every } V_n \textit{ real});$$

each box has an extra 'potential' V_n associated with it. The limit $\delta \to 0$ gives in this case

$$i\hbar \frac{\partial \psi}{\partial t}(x, t) = -\frac{\hbar^2}{2m}\frac{\partial^2 \psi(x, t)}{\partial x^2} + V(x)\psi(x, t);$$

this is the Schrödinger equation for a particle moving in an external real potential $V(x)$. The Hamiltonian for this system is

$$H = -\frac{\hbar^2}{2m}\frac{\partial^2}{\partial x^2} + V(x),$$

the sum of two linear operators, one a differential operator and the other a simple operator of multiplication by $V(x)$.

PROBLEM

Show that the complex conjugate wavefunction $\psi^*(x, t)$ (which incidentally corresponds to the Hermitian conjugate row vector with components c_n^*) satisfies

$$-i\hbar\frac{\partial \psi^*}{\partial t} = -\frac{\hbar^2}{2m}\frac{\partial^2 \psi^*}{\partial x^2} + V\psi^*.$$

Mean values

Earlier it was shown (p. 18) that the mean value of a (matrix) observable X in a state with state vector u may be written $u^+ Xu$. We need an analogous expression for the new kind of operator with which we shall now be concerned.

An operator—at least for the moment—is an object which transforms one function into another; if we write

$$\phi(x) = O\psi(x)$$

we mean that the effect of O on the function $\psi(x)$ is to produce the function $\phi(x)$. A **linear operator** has a further 'distributive' property:

$$O(\lambda\psi_1(x)+\mu\psi_2(x)) = \lambda O\psi_1(x)+\mu O\psi_2(x)$$

(one may 'multiply out brackets'). We may distinguish several different types of linear operator, but we are interested here in only two kinds:

(1) *Multiplicative:* $O = f(x)$, i.e. $\phi(x) = f(x)\psi(x)$. ϕ is obtained from ψ simply by multiplying by $f(x)$.

(2) *Derivative:* an example is $O = \mathrm{d}/\mathrm{d}x$, i.e. $\phi(x) = \mathrm{d}\psi(x)/\mathrm{d}x$; ϕ is obtained from ψ by differentiating with respect to x.

We may get a prescription for the mean value of any type of linear operator as follows. Consider first the familiar matrix analogue $u^+ Xu$ (p. 18), in a slightly different way: apply the matrix X to the vector u to produce a new vector $v = Xu$, and then form the scalar product $u^+ v$. This prescription may be immediately adapted to our present needs:

(a) apply the operator O to the function $\psi(x)$ to produce a new function $\phi(x)$;

(b) form the scalar product (see p. 35)

$$\int_{-\infty}^{\infty} \psi^*(x)\phi(x)\,\mathrm{d}x.$$

Taking steps (a) and (b) together yields an expression for the mean value of O in the state with wavefunction $\psi(x)$,

$$\langle O\rangle_\psi = \int_{-\infty}^{\infty} \psi^*(x)\,.\,O\psi(x)\,\mathrm{d}x.$$

We shall use expressions of this type repeatedly in the next few pages as we use the correspondence principle to find the operators which represent the observables of familiar entities like position and momentum.

Position and velocity observables

Consider the particle to be in the state represented by the normalized wavefunction $\psi(x)$. The probability that an observation of the position of the particle yields a result in the range $(x, x+dx)$ is $|\psi(x)|^2\,dx$. Therefore, in the usual sense of statistics,

$$\text{the mean position of the particle} \equiv \langle x\rangle_\psi$$

$$= \int_{-\infty}^{\infty} x|\psi(x)|^2\,dx = \int_{-\infty}^{\infty} \psi^*(x)\,.\,x\psi(x)\,dx.$$

(The integral is assumed to converge.) The right side is of the form of the mean value of the multiplication operator x. In line with the work of the last section, we therefore say

the observable x is the multiplication operator x.

To find an observable to represent the *velocity* v of the particle, we use the correspondence principle in the form: the time derivative of $\langle x\rangle_\psi$ is the *mean* velocity $\langle v\rangle_\psi$. That is,

$$\langle v\rangle_\psi = \frac{d}{dt}\int_{-\infty}^{\infty} \psi^* x\psi\,dx$$

$$= \int_{-\infty}^{\infty}\left(\frac{\partial\psi^*}{\partial t}x\psi + \psi^* x\frac{\partial\psi}{\partial t}\right)dx$$

$$= \frac{1}{i\hbar}\int_{-\infty}^{\infty}\left\{\left(\frac{\hbar^2}{2m}\frac{\partial^2\psi^*}{\partial x^2} - V\psi^*\right)x\psi + \psi^* x\left(-\frac{\hbar^2}{2m}\frac{\partial^2\psi}{\partial x^2} + V\psi\right)\right\}dx$$

$$= \int_{-\infty}^{\infty}\psi^*\left(-\frac{i\hbar}{m}\frac{\partial\psi}{\partial x}\right)dx.$$

(The last step requires several integrations by parts, assuming that the integrated parts at $\pm\infty$ vanish.) The right side is the mean value of the derivative operator $(-i\hbar/m)\partial/\partial x$, and we therefore say

the observable v is the derivative operator $-\dfrac{i\hbar}{m}\dfrac{\partial}{\partial x}$.

Newcomers to this type of result are often surprised that velocity, normally thought of as a derivative with respect to t, should also be connected with a derivative with respect to x. However, the conclusion has been reached by a route which can hardly be avoided if the quantum-mechanical approach is going to give Newton's laws for mean values.

PROBLEMS

1. Show that $\psi(x) = A(x^2+a^2)^{-\frac{1}{2}}$ may be normalized by choosing the constant A correctly, but that $\langle x\rangle_\psi$ does not exist for this wavefunction.

2. Show that $\psi(x) = B(x^2+a^2)^{-1}\exp(\mathrm{i}x^4)$ may be normalized, but that $\langle v\rangle_\psi$ does not exist for this wavefunction. (The point of these two examples is that for a *physical* state we require more of a wavefunction than that it be merely normalized.)
3. Assuming that $\langle v\rangle_\psi$ exists, show that it is real only if $|\psi(x)| \to 0$ as $x \to$ either $+\infty$ or $-\infty$. (Moral: now that we are dealing with differential operators, *boundary conditions have become important.*)
4. Carry out explicitly the integrations by parts in the last section.
5. Show that for any function $\psi(x)$,

$$\left(-\frac{\hbar^2}{2m}\frac{\partial^2}{\partial x^2}+V(x)\right)x\psi = x\left(-\frac{\hbar^2}{2m}\frac{\partial^2}{\partial x^2}+V(x)\right)\psi-\frac{\hbar^2}{m}\frac{\partial}{\partial x}\psi.$$

Deduce that $[x, H] = \mathrm{i}\hbar v$, and explain the relevance of this to the last section.

Momentum, force, and energy

Let us go on to consider the mean *acceleration* of the particle in the state ψ,

$$\begin{aligned}\frac{\mathrm{d}}{\mathrm{d}t}\langle v\rangle_\psi &= \frac{\mathrm{d}}{\mathrm{d}t}\int_{-\infty}^{\infty}\psi^*\left(-\frac{\mathrm{i}\hbar}{m}\frac{\partial\psi}{\partial x}\right)\mathrm{d}x\\ &= -\frac{\mathrm{i}\hbar}{m}\int_{-\infty}^{\infty}\left(\frac{\partial\psi^*}{\partial t}\frac{\partial\psi}{\partial x}+\psi^*\frac{\partial^2\psi}{\partial x\partial t}\right)\mathrm{d}x\\ &= -\frac{1}{m}\int_{-\infty}^{\infty}\left\{\left(\frac{\hbar^2}{2m}\frac{\partial^2\psi^*}{\partial x^2}-V\psi^*\right)\frac{\partial\psi}{\partial x}+\psi^*\frac{\partial}{\partial x}\left(-\frac{\hbar^2}{2m}\frac{\partial^2\psi}{\partial x^2}+V\psi\right)\right\}\mathrm{d}x\end{aligned}$$

on using Schrödinger's equation to substitute for $\partial\psi/\partial t$ and $\partial\psi^*/\partial t$. After several integrations by parts, and assuming that all integrated parts vanish, we find

$$m\frac{\mathrm{d}}{\mathrm{d}t}\langle v\rangle_\psi = \int_{-\infty}^{\infty}\psi^*\left(-\frac{\partial V}{\partial x}\right)\psi\,\mathrm{d}x = \left\langle-\frac{\partial V}{\partial x}\right\rangle_\psi.$$

This is exactly Newton's equation of motion of a particle in the form

mass × mean acceleration = mean force,

provided that certain further identifications are made:

classical concept	observable	
mass	m	a constant
force	$-\dfrac{\partial V}{\partial x}$	a multiplication operator.

Other natural identifications follow immediately:

momentum (mv) $\quad -i\hbar\dfrac{\partial}{\partial x}$

potential energy (V) $\quad V(x) \quad \left(\text{since force} = -\dfrac{\partial V}{\partial x}\right)$

kinetic energy ($\frac{1}{2}mv^2$) $\quad -\dfrac{\hbar^2}{2m}\dfrac{\partial^2}{\partial x^2}$

total energy $\quad -\dfrac{\hbar^2}{2m}\dfrac{\partial^2}{\partial x^2}+V(x) \quad$ (the Hamiltonian!).

To summarize, it has been shown that adopting Schrödinger's equation as the proper description of a quantum particle allows us to identify unambiguously certain observables whose mean values obey the equations of motion for a classical particle. This provides the first step towards showing that macroscopic bodies (even though composed of *quantum* particles) will satisfy classical Newtonian equations of motion, in agreement with experimental observation. A complete demonstration is beyond the scope of this book.

However, it is worth emphasizing that this behaviour of mean values is much more compelling as a reason for adopting Schrödinger's equation for a quantum particle than our original arguments of pp. 34–37. These arguments depended on an idea that 'action at a distance' is in some way unsatisfactory, which in its turn restricted the equation of motion drastically. But why *should* 'action at a distance' be unsatisfactory? The ultimate appeal must always be to the results of experiment; in particular, macroscopic bodies obey Newton's laws, and this fact carries much weight.

PROBLEM

Show that $[mv, H] = -i\hbar(\partial V/\partial x)$, and explain the relevance of this to the last section.

The eigenvalue problem

The possible results of measuring an observable are the eigenvalues of that observable. The eigenvalues of the particle position x therefore comprise any real value whatever. Allowing eigenvalues to belong to a continuous range in this way brings new problems which are mathematical, not physical, in nature. For example, the eigenvalue equation for the operator x with eigenvalue x_0 is

$$x\psi_{x_0}(x) = x_0\psi_{x_0}(x), \qquad \text{i.e. } (x-x_0)\psi_{x_0}(x) = 0.$$

This implies that $\psi_{x_0}(x)$ (as a function of x) is zero everywhere except at $x = x_0$! This function will hardly do to represent any kind of *physical* state.

Actually, this is not really a difficulty, since the infinite precision required to get a particle into such a state (of being *exactly* at the point x_0) is not attainable in practice. Nevertheless, we still insist that x possesses eigenvalues in a continuous range, even though the eigenstates are physically unattainable, because it is much more convenient to do so.

The same kind of thing happens with the observable for velocity $v = (-i\hbar/m)\partial/\partial x$, though the details are different. The eigenvalue equation for v is

$$-\frac{i\hbar}{m}\frac{\partial\psi_{v_0}(x)}{\partial x} = v_0\psi_{v_0}(x),$$

and this has a solution for any number v_0, real or complex, namely, $\psi_{v_0}(x) = \exp(imv_0x/\hbar)$. For physical sense, only real values of v_0 are allowed; non-real values may be excluded by requiring ψ to be *bounded* at ∞. (In general it is true that the eigenvalue problem for a differential operator must always specify the boundary conditions.) Even so, if v_0 is real, $|\psi_{v_0}(x)| = 1$ for all x and ψ cannot be normalized. Thus eigenstates of v are not real physical states; the possible values of v form a continuous range, and infinite precision would be needed to achieve an eigenstate.

PROBLEMS

1. Show that the state $\psi_a(x) = (\pi a)^{-\frac{1}{4}} \exp(-x^2/2a)$ is normalized, provided $a > 0$, and evaluate the mean value of x and the uncertainty in x for this state. $(0, \sqrt{(a/2)})$. Describe how this state changes as $a \to 0$. (You will need

$$\langle x^2 \rangle_\psi = \int_{-\infty}^{\infty} \psi^*(x)x^2\psi(x)\,dx.)$$

2. For the state $\psi_a(x)$ of problem 1, show that the mean value of the Hamiltonian $(-\hbar^2/2m)\partial^2/\partial x^2$ is $\hbar^2/4am$. What can be said as $a \to 0$?

3. In the light of the last two problems, can you suggest a possible experimental difficulty in getting a quantum particle (say an electron) into a small volume (say the size of a hydrogen atom)?

4. Carry out explicitly the integrations by parts in the last section but one.

The stationary states of a free particle

The eigenstates of v are also stationary states for the free particle, since

$$-\frac{\hbar^2}{2m}\frac{\partial^2}{\partial x^2}e^{ikx} = \frac{\hbar^2k^2}{2m}e^{ikx}.$$

Consequently, we say that the (time-dependent) wavefunction for a particle of mass m, velocity $\hbar k/m$, momentum $\hbar k$, and kinetic energy $\hbar^2k^2/2m$ is

$$\psi_k(x, t) = e^{ikx - i\omega t}, \qquad \text{with } \hbar\omega = \hbar^2k^2/2m.$$

Even though such a state is not normalizable, we often use it in the following way: imagine a *beam* of particles with no mutual interaction, all with the same velocity, momentum, and kinetic energy, the particle density of the beam being ρ particles per unit volume. Then we shall say that the wavefunction describing the beam is

$$\rho^{\frac{1}{2}}\, e^{ikx-i\omega t},$$

the particle current being $\rho\hbar k/m$. This slightly improper interpretation relies on the possibility of regarding the beam as a very large number of independent 'experiments' conducted for convenience at the same place and time; in fact, this idea is very close to the usual practice in scattering experiments. It is important that the interaction between the particles in the beam should be negligible for this approach to be valid, otherwise a full quantum theory of many particles will be needed.

These ideas may be extended to beams of particles with more than one velocity. A very common case in elementary applications is given by the wavefunction

$$\psi(x, t) = A\, e^{ikx-i\omega t} + B\, e^{-ikx-i\omega t}.$$

This is an eigenfunction of the Hamiltonian (and therefore a stationary state), but it is *not* an eigenfunction of velocity or momentum; it is a *superposition* of two such eigenfunctions. This wavefunction represents a beam of particles with the property that measuring the velocity of a particle picked at random will yield $\hbar k/m$ with probability $|A|^2/(|A|^2+|B|^2)$, and $-\hbar k/m$ with probability $|B|^2/(|A|^2+|B|^2)$; i.e. the current of particles to the 'right' is $|A|^2\hbar k/m$, and to the 'left' $-|B|^2\hbar k/m$. Such a wavefunction is useful in problems involving the *reflection* of particles from a potential barrier.

PROBLEMS

1. If $\psi(x, t)$ satisfies the Schrödinger equation with potential $V(x)$, show that the *continuity equation*

$$\frac{\partial|\psi|^2}{\partial t} + \frac{\partial s(x, t)}{\partial x} = 0$$

is satisfied, where

$$s(x, t) = -\frac{i\hbar}{2m}\left(\psi^*\frac{\partial\psi}{\partial x} - \psi\frac{\partial\psi^*}{\partial x}\right).$$

(Since $|\psi|^2$ is a particle density—in the interpretation in terms of a beam of particles—the continuity equation shows that s is a **particle current.** This may be seen by integrating the continuity equation with respect to x,

$$\frac{d}{dt}\int_{x_1}^{x_2} |\psi|^2\, dx = s(x_1, t) - s(x_2, t),$$

with the interpretation

rate of change of particle number in segment $x_1 \leqslant x \leqslant x_2$
= flow in at x_1 − flow out at x_2.)

2. For the stationary state $\psi = A\exp\{i(kx-\omega t)\}+B\exp\{i(-kx-\omega t)\}$, show that $s = |A|^2\hbar k/m - |B|^2\hbar k/m$, and interpret the two terms. Show also that

$$|\psi|^2 = |A|^2+|B|^2+AB^*\,e^{2ikx}+BA^*\,e^{-2ikx}.$$

(In evaluating s, the cross-terms cancel. This is not true of $|\psi|^2$, which fluctuates about an average value $|A|^2+|B|^2$. Thus a particle in such a beam is *more likely* to be found at some places than at others. This **interference** is typical of superposed wave motions; it was with this in mind that the earlier quantum theorists described quantum particles as behaving 'sometimes like particles, sometimes like waves'. This was how the wavefunction received its name, while Schrödinger's original approach to quantum mechanics was called *wave mechanics*.)

The Heisenberg uncertainty principle

By the ordinary formula for the differentiation of a product,

$$\frac{d}{dx}\{x\psi(x)\} = x\frac{d\psi}{dx}+\psi.$$

In terms of the observable for linear momentum p $(= -i\hbar\partial/\partial x)$, this may be written

$$px\psi = xp\psi - i\hbar\psi$$

or $[x, p]\psi = i\hbar\psi$. Since this is true for any wavefunction $\psi(x)$, it follows that the commutator

$$[x, p] = i\hbar,$$

a particularly simple result.

We may now apply the work on uncertainty relations (p. 24) as follows: suppose that for a particular state the uncertainties in x and p are respectively Δ_x and Δ_p; then

$$\Delta_x\Delta_p \geqslant \tfrac{1}{2}|\langle[x, p]\rangle| = \tfrac{1}{2}\hbar$$

(note that the mean value of a constant is just that constant itself). This is the celebrated **Heisenberg uncertainty principle**; it imposes an *absolute limit* on the accuracy with which position and momentum may be simultaneously measured.

PROBLEM

Show that for the state $\psi_a(x)$ of problems 1 and 2 (p. 42), $\Delta_x\Delta_p$ is actually *equal* to $\frac{1}{2}\hbar$. (These problems have already given $\langle x\rangle = 0$, $\langle x^2\rangle = \frac{1}{2}a$, and $\langle p^2\rangle = \hbar^2/2a$, and it is simple to show that $\langle p\rangle = 0$. It remains to apply the general formula on p. 23.)

5. Quantum particles in one dimension: some examples

Energy spectra

WE have seen that the Hamiltonian is the observable for the energy of a system. The set of all eigenvalues of the Hamiltonian is therefore of central interest, particularly when the Hamiltonian itself does not depend on time; the set of energy eigenvalues is the **energy spectrum** of the system. ('Spectrum' is a much overworked word; its use here should be carefully distinguished from a closely related use—the emission or absorption 'spectrum' of an atom or molecule.)

There are very many different ways in which the energy eigenvalues may be distributed. The examples in this chapter have been chosen to illustrate some of the more important types of spectra from the physical point of view; they by no means exhaust all the possibilities, but they are fairly representative of some 'real-life' situations.

The free particle has already been considered. The energy eigenvalues ($\hbar^2k^2/2m$) form a continuous range from zero to $+\infty$. Confining the particle

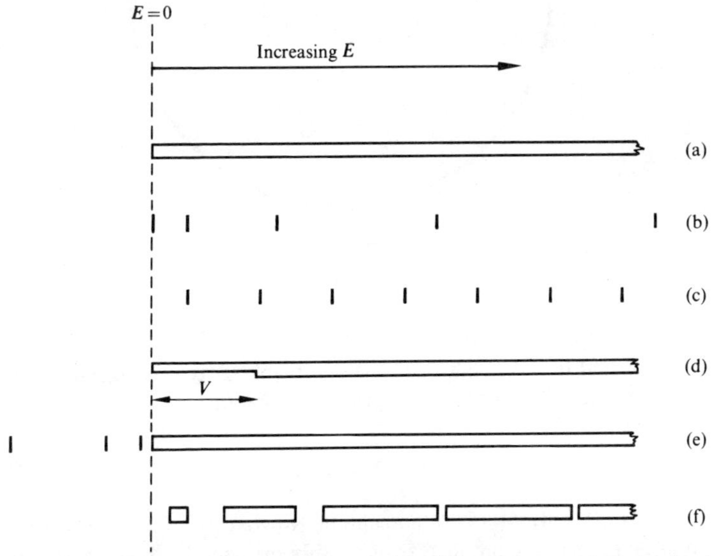

Typical energy eigenvalue spectra: (a) free particle; (b) particle on a circle; (c) simple harmonic oscillator; (d) potential step; (e) potential well; (f) Krönig–Penney model.

to a finite region or introducing a potential $V(x)$ may modify the eigenvalue spectrum in several ways. The spectra for the different systems we shall consider are sketched on one figure for convenient comparison.

A particle moving on a circle

A quantum particle of mass m is constrained to move on a circle of radius a; its position on the circumference is conveniently specified by an angle θ. The coordinate which corresponds to what was x is $s = a\theta$, the distance measured round the circumference. The observable for linear momentum along the circumference is thus $-i\hbar\partial/\partial s$.

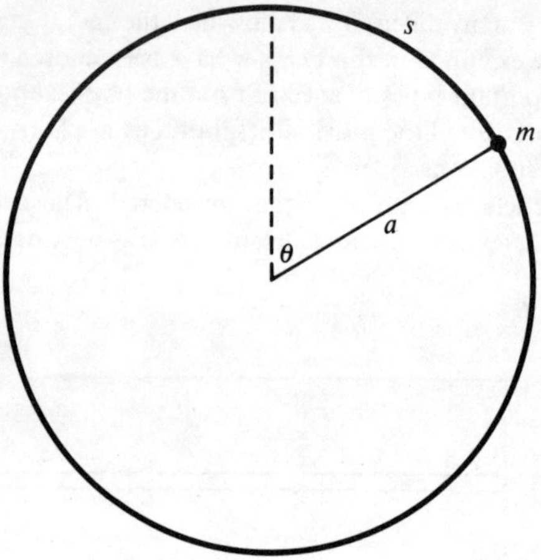

Let us find the eigenvalues of this observable. A typical **eigenfunction** will be a function $\psi(\theta)$ satisfying the eigenvalue equation

$$-i\hbar\frac{\partial\psi}{\partial s} \equiv -\frac{i\hbar}{a}\frac{\partial\psi}{\partial\theta} = p\psi,$$

where p is an eigenvalue of the momentum. Now, an obvious but fundamentally important remark is that increasing θ by 2π brings us back to the *same* point on the circle; consequently ψ must be *periodic with period* 2π. The eigenvalues and eigenfunctions of momentum are therefore

$$p_n = \hbar n/a, \qquad \psi_n(\theta) = (2\pi)^{-\frac{1}{2}}\, e^{in\theta},$$

where, because of the periodicity condition, n is any positive or negative integer, including zero:

$$n = \ldots -3, -2, -1, 0, 1, 2, \ldots .$$

(For the reason for the normalization factor $(2\pi)^{-\frac{1}{2}}$, see problem 1, below).

In the absence of external forces, the Hamiltonian for this system is

$$H = -\frac{\hbar^2}{2m}\frac{\partial^2}{\partial s^2} = -\frac{\hbar^2}{2ma^2}\frac{\partial^2}{\partial \theta^2}.$$

Its eigenvalues are the possible results of measuring kinetic energy, and they are $E_n = p_n^2/2m = \hbar^2 n^2/2ma^2$. Since $E_n = E_{-n}$, these eigenvalues are doubly degenerate, except for E_0.

The eigenvalue spectrum of H is discrete in this example, illustrating a general result: *a particle confined to a finite volume will have a discrete energy spectrum.* The individual eigenvalues of a discrete energy spectrum are called **energy levels.**

PROBLEMS

1. Show that the functions ψ_n are orthonormal in the sense

$$(\psi_m^*, \psi_n) = \int_0^{2\pi} \psi_m^* \psi_n \, d\theta = \delta_{mn}.$$

2. Show that the functions $\cos n\theta$ and $\sin n\theta$ are eigenfunctions of the Hamiltonian, but not of the momentum. What has this result to do with degeneracy?

Angular momentum

For the particle moving on a circle (considered in the last section) the *angular momentum* about the centre is $a \times$(momentum round the circumference). Thus the observable for this angular momentum is $-i\hbar\partial/\partial\theta$, with eigenvalues $ap_n = \hbar n$, with n any integer. Thus $\hbar$ is a natural unit for angular momentum in this case. (In fact, it may be shown that this is universally true.)

We may distinguish two possible types of angular momentum. **Orbital angular momentum** is associated with an angular position coordinate (like θ) and a periodicity of 2π in the wavefunction; this type always has eigenvalues of the form $n\hbar$, where n is *any integer whatever*. The other type is **intrinsic angular momentum** (or **spin**) and is *never* associated with an angular coordinate. Photon angular momentum is partly intrinsic, and the intrinsic part takes just two possible values $\pm\hbar$. It would be wrong to regard this as arising through some curious trajectory of the photon; it is an intrinsic property of the photon itself.

In Chapter 7 we shall see that the electron is another particle which possesses a spin. The general theory of angular momentum is beyond the scope of this book.

The simple harmonic oscillator

A classical particle of mass m moving on the x-axis and subject to a restoring force $-kx$ has a total energy $\frac{1}{2}m\dot{x}^2+\frac{1}{2}kx^2$; and it will execute a *simple harmonic motion* of angular frequency $\omega = \sqrt{(k/m)}$. The corresponding quantum system has the Hamiltonian

$$H = -\frac{\hbar^2}{2m}\frac{\partial^2}{\partial x^2}+\tfrac{1}{2}kx^2,$$

and by the correspondence principle the *mean value* of x in any state will execute a simple harmonic motion of angular frequency ω.

By the general work on the time-variation of mean values (p. 33), we infer that the eigenvalues of the Hamiltonian must be separated by gaps of $\hbar\omega$ (if the gaps were less, the mean value of x would be expected to have oscillatory components with frequency less than ω). The quantum simple harmonic oscillator therefore has a *discrete* energy spectrum.

It is always true that if the potential $V(x) \to +\infty$ as $x \to +\infty$ and as $x \to -\infty$ the energy spectrum is discrete, never continuous. The simple harmonic oscillator is the special case of $V(x) = \frac{1}{2}kx^2$.

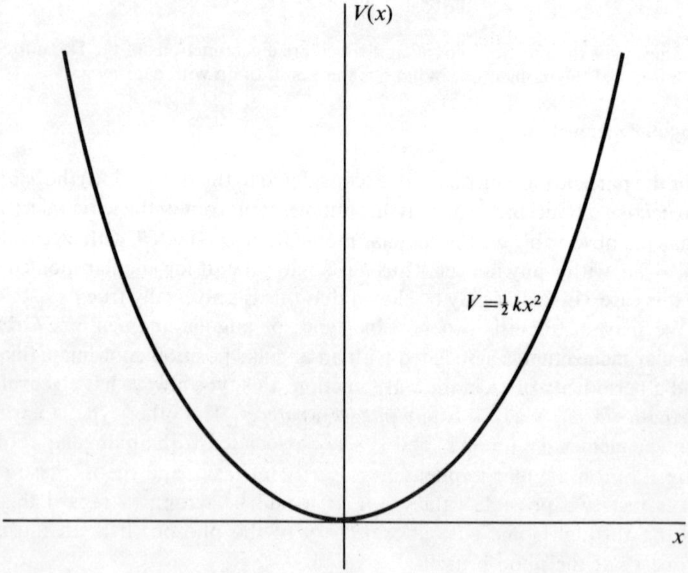

The importance of the simple harmonic oscillator as a quantum system cannot be overemphasized, particularly for its application to the electromagnetic field. Its complete solution is undertaken in chapter 11, and we shall merely mention here that its energy spectrum is the sequence of values $\hbar\omega(n+\frac{1}{2})$, with $n = 0, 1, 2, \ldots$.

PROBLEMS

1. Show that, provided $\alpha(>0)$ is correctly chosen, $\exp(-\alpha x^2)$ and $x\exp(-\alpha x^2)$ are both eigenfunctions of the Hamiltonian for the harmonic oscillator, with eigenvalues $\frac{1}{2}\hbar\omega$ and $\frac{3}{2}\hbar\omega$ respectively. (Substitute into the Schrödinger equation, and equate coefficients of different powers of x to zero.)
2. Show that, for a proper choice of A, $(x^2+A)\exp(-\alpha x^2)$ is an eigenfunction of the Hamiltonian, with eigenvalue $\frac{5}{2}\hbar\omega$.
3. Show that the three eigenfunctions in the last two problems are mutually orthogonal in pairs.
4. If x, v, and H are the observables for position, velocity, and energy, show that

$$(i\hbar)^{-1}[x, H] = v \quad \text{and} \quad (i\hbar)^{-1}[v, H] = -\omega^2 x.$$

Hence show that the mean value $\langle x\rangle$ in any state satisfies the classical simple harmonic equation of motion

$$\frac{d^2\langle x\rangle}{dt^2}+\omega^2\langle x\rangle = 0.$$

The potential step

A system of a quite different character is obtained by taking

$$V(x) = \begin{cases} 0 & (x < 0), \\ V\,(\text{constant}) & (x > 0). \end{cases}$$

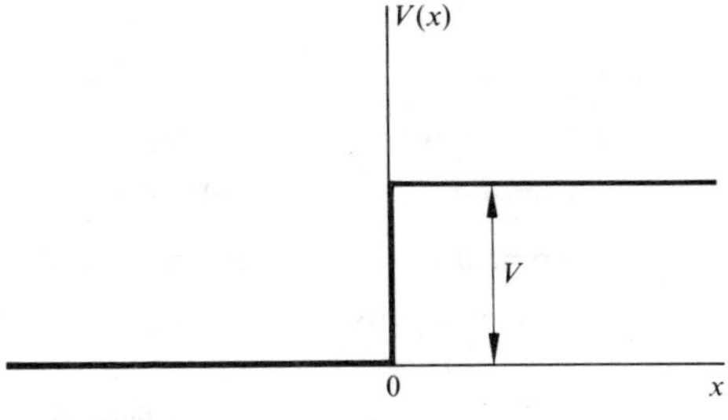

The Schrödinger equation for a *stationary* state of energy E now takes two different forms according to the value of x:

$$\text{for } x < 0: \quad -\frac{\hbar^2}{2m}\frac{\partial^2\psi}{\partial x^2} = E\psi,$$

$$\text{for } x > 0: \quad -\frac{\hbar^2}{2m}\frac{\partial^2\psi}{\partial x^2} + V\psi = E\psi.$$

At the 'awkward' point $x = 0$ we adopt the boundary conditions

$$\psi(x) \text{ and } \frac{\partial\psi}{\partial x} \text{ are both continuous at } x = 0;$$

this is sufficient to ensure that particle current (p. 43) is continuous at the boundary. These conditions represent the most natural choice anyway.

The case $E > V$

For any such E, we may immediately write down the two parts of the most general solution of the Schrödinger equation:

$$\psi(x) = \begin{cases} A\exp(ik_1x) + B\exp(-ik_1x) & \text{for } x < 0, \\ C\exp(ik_2x) + D\exp(-ik_2x) & \text{for } x > 0, \end{cases}$$

where the (positive) constants k_1 and k_2 are defined by

$$\hbar^2k_1^2 = 2mE \quad \text{and} \quad \hbar^2k_2^2 = 2m(E - V).$$

The constant coefficients A, B, C, and D are not independent but are related by the continuity requirements at $x = 0$:

$$A + B = C + D \quad \text{and} \quad ik_1(A - B) = ik_2(C - D).$$

These two relations reduce the four coefficients to *two* independent ones.

According to the discussion on p. 43, the resulting solution may be interpreted in the following way; there are four beams of particles, as follows,

	$x < 0$	$x > 0$
left to right:	momentum k_1, current $k_1\lvert A\rvert^2$	momentum k_2, current $k_2\lvert C\rvert^2$
right to left:	momentum $-k_1$, current $-k_1\lvert B\rvert^2$	momentum $-k_2$, current $-k_2\lvert D\rvert^2$.

A reasonable 'experimental' application is to the case of a beam from the left (A) being partially transmitted (C) and partially reflected (B) at the step. In this case $D = 0$, and we may obtain B and C in terms of A:

$$B = \frac{k_1 - k_2}{k_1 + k_2}A \quad \text{and} \quad C = \frac{2k_1}{k_1 + k_2}A.$$

Hence

$$k_1|B|^2 = R\,.\,k_1|A|^2 \quad \text{and} \quad k_2|C|^2 = T\,.\,k_1|A|^2,$$

with **reflection** and **transmission coefficients**

$$R = \frac{(k_1 - k_2)^2}{(k_1 + k_2)^2} \quad \text{and} \quad T = \frac{4k_1 k_2}{(k_1 + k_2)^2}$$

with, inevitably, $R + T = 1$.

As is usual in 'scattering' problems of this kind, the energy spectrum is continuous, there being states for each $E > V$. Since the general state for fixed $E(> V)$ depends on *two* parameters, the eigenvalue E is twofold **degenerate** (see p. 15). If $0 \leqslant E \leqslant V$, the corresponding state is non-degenerate. There are no eigenvalues less than 0. (See problems below).

PROBLEMS

1. Show that an eigenfunction for an eigenvalue E *between* 0 and V must take the form

$$\psi(x) = \begin{cases} A\,\mathrm{e}^{ikx} + B\,\mathrm{e}^{-ikx} & \text{for } x < 0, \\ F\,\mathrm{e}^{-\lambda x} & \text{for } x > 0, \end{cases}$$

with

$$\hbar^2 k^2 = 2mE, \qquad \hbar^2 \lambda^2 = 2m(V - E), \qquad A + B = F \quad \text{and} \quad ik(A - B) = -\lambda F.$$

Show that the particle current (p. 43) is zero everywhere, and deduce that $R = 1$, $T = 0$. (This is the case of **total reflection**: any particle coming towards the step from the left will *eventually* be found moving back to the left. 'Eventually', because the reversal of direction is not sudden; quantum barriers are 'spongy', in the sense that quantum particles may penetrate them in a way that classical particles may not. For $x > 0$, $|\psi|^2 = |F|^2 \exp(-2\lambda x)$; thus there is a non-zero probability of finding the particle anywhere to the *right* of the step, though this probability becomes very small as x becomes large.)

2. Show that, if it existed, an eigenfunction for a value $E < 0$ would have to take the form

$$\psi(x) = \begin{cases} G \exp(\lambda_1 x) & (x < 0), \\ H \exp(-\lambda_2 x) & (x > 0), \end{cases}$$

with

$$\hbar^2 \lambda_1^2 = 2m|E|, \qquad \hbar^2 \lambda_2^2 = 2m(V - E),$$

but that it is impossible to satisfy the continuity requirements.

3. Show that the eigenfunction in problem 1 depends on only one disposable parameter, apart from E. Deduce that it is non-degenerate.

The potential well

The potential well may be described as two potential steps facing each other; it is defined by

$$V(x) = \begin{cases} 0 & (x < 0) \\ V\,(\text{negative}) & (0 < x < a), \\ 0 & (a < x). \end{cases}$$

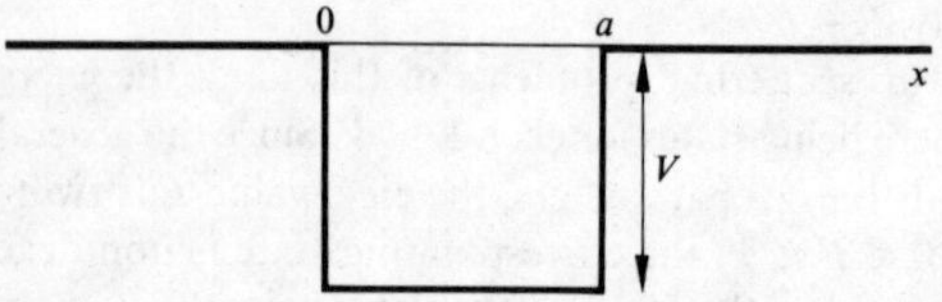

The Schrödinger equation takes different forms according to whether x is 'in' the well or not:

$$\text{for } 0 < x < a\text{:}\quad -\frac{\hbar^2}{2m}\frac{\partial^2\psi}{\partial x^2} + V\psi = E\psi;$$

$$\text{for }\; x < 0 \;\text{ and }\; \text{for }\; a < x\text{:}\quad -\frac{\hbar^2}{2m}\frac{\partial^2\psi}{\partial x^2} = E\psi;$$

consequently it is necessary to write down the wavefunction in *three* pieces.

The case $V < E < 0$

The solution in the three regions takes the form

$$\psi(x) = \begin{cases} A\,e^{\lambda x} & (x < 0), \\ B\,e^{ikx} + C\,e^{-ikx} & (0 < x < a), \\ D\,e^{-\lambda x} & (a < x), \end{cases}$$

with k and λ defined by

$$\hbar^2 k^2 = 2m(E - V) \quad\text{and}\quad \hbar^2\lambda^2 = -2mE.$$

Again we have continuity requirements at the boundaries: ψ and $\partial\psi/\partial x$ must both be continuous at both $x = 0$ and $x = a$. This gives four conditions

	$\underline{x = 0}$	$\underline{x = a}$
ψ :	$A = B + C$	$B\,e^{ika} + C\,e^{-ika} = D\,e^{-\lambda a}$
$\partial\psi/\partial x$:	$\lambda A = ik(B - C)$	$ik(B\,e^{ika} - C\,e^{-ika}) = -\lambda D\,e^{-\lambda a}$.

These are in fact four linear homogeneous equations linking the four coefficients A, B, C, and D. In order to have a physically meaningful solution, not all of A, B, C, and D must be zero; this is possible only if the determinant of the equations vanishes:

$$\begin{vmatrix} 1 & -1 & -1 & \cdot \\ \lambda & -ik & ik & \cdot \\ \cdot & e^{ika} & e^{-ika} & -e^{-\lambda a} \\ \cdot & ik\,e^{ika} & -ik\,e^{-ika} & \lambda\,e^{-\lambda a} \end{vmatrix} = 0.$$

After some manipulation, this requirement reduces to

$$(k^2-\lambda^2)\tan ka = 2k\lambda.$$

Substituting for k and λ now gives

$$\tan\left[\frac{a}{\hbar}\sqrt{\{2m(E-V)\}}\right] = 2\sqrt{\{E(V-E)\}}/(2E-V),$$

where the positive square roots must be taken. The roots of this equation give the allowed values of E in the range $V < E < 0$; the spectrum is thus *discrete* in this range.

The other cases are referred to in the problems. For $0 < E$ the spectrum is continuous, and we may obtain transmission and reflection coefficients as before. There are no eigenvalues in the range $E < V(<0)$. This example illustrates the possibility of a spectrum that is partly continuous and partly discrete. This may happen for any potential which is zero at large distances and takes negative values in some regions. The discrete eigenvalues give the **bound states**, while the continuous range gives the **scattering states**.

PROBLEMS

1. Evaluate the determinant on this page.

2. Show that 'scaling' E and a by the prescription

$$E = -\varepsilon V \quad \text{and} \quad \hbar\alpha = a\sqrt{(-2mV)},$$

leads to the requirement

$$\tan\{\alpha\sqrt{(1+\varepsilon)}\} = -2\sqrt{\{-\varepsilon(1+\varepsilon)\}}/(2\varepsilon+1),$$

in the range $-1 \leq \varepsilon \leq 0$. (This kind of scaling is often useful for discussing problems containing a number of parameters. In this case, the problem depends on two parameters V and a, but the essential features depend only on the *combination* Va^2 or, equivalently, α.)

3. Show that if $(n-1)\pi < \alpha < n\pi$ the system has just n bound states. (As α increases, new discrete eigenvalues detach themselves from the bottom ($\varepsilon = 0$) of the continuous range. Carry out a graphical solution.)

4. *The case* $E > 0$. Show that it is possible to have a solution of the type

$$\psi(x) = \begin{cases} A\exp(ik_1x) + B\exp(-ik_1x) & (x < 0), \\ C\exp(ik_2x) + D\exp(-ik_2x) & (0 < x < a), \\ F\exp(ik_1x) & (a < x), \end{cases}$$

and interpret the various terms. Write down the four continuity requirements and hence obtain the transmission and reflection coefficients

$$T = k_1|F|^2/k_1|A|^2 \quad \text{and} \quad R = k_1|B|^2/k_1|A|^2.$$

Show that for energies giving $k_2 = n\pi/a$, there is *perfect transmission* ($T = 1$, $R = 0$). (A similar phenomenon is well known in the optics of thin films, where it is used, for example, in the 'blooming' of camera lenses. It is essentially an interference effect characteristic of wave motion.)

The Krönig–Penney model

This was proposed in 1930 as a model for a 'one-dimensional crystal', and it correctly predicts the general features of the spectrum of a particle moving in any *periodic* potential.

The potential is a periodic arrangement of potential wells of depth V and width b, separated by plateaux of width a:

$$V(x) = \begin{cases} 0 & n(a+b) < x < n(a+b)+b \quad \text{(wells)}, \\ V\,\text{(constant)} & n(a+b)+b < x < (n+1)(a+b) \quad \text{(plateaux)}, \end{cases}$$

with n each integer in turn. $V(x)$ is periodic with period $l = a+b$.

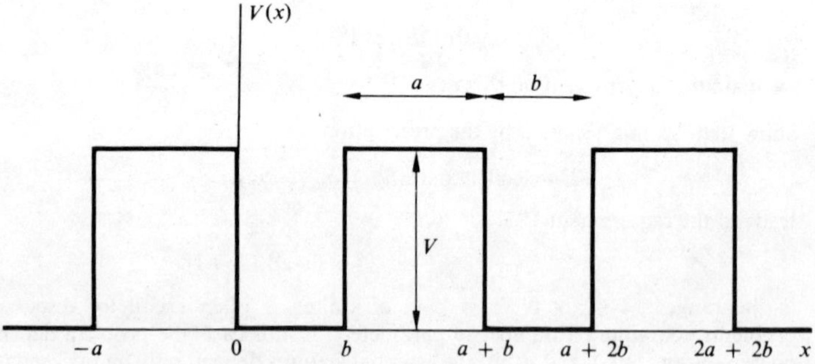

We shall search for stationary states of a rather special kind; the wavefunctions are to satisfy a kind of periodicity condition related to the periodicity of $V(x)$. We require $\psi(x)$ to be *periodic apart from a phase factor*

$$\psi(x+l) = e^{iKl}\psi(x)$$

where K is a real constant. It was shown by Bloch in 1928 that any state may be written as a linear combination of solutions of this kind, and so restricting our search in this way brings no loss of generality. It is now sufficient to consider ψ in one well and on one adjacent plateau; the periodicity condition will give ψ at all other points.

Thus we take (for the case $E > V$)

$$\psi(x) = \begin{cases} A\exp(ik_1x)+B\exp(-ik_1x) & (-a < x < 0) \quad \text{(plateau)}, \\ C\exp(ik_2x)+D\exp(-ik_2x) & (0 < x < b) \quad \text{(well)}, \end{cases}$$

where

$$\hbar^2k_1^2 = 2m(E-V) \quad \text{and} \quad \hbar^2k_2^2 = 2mE.$$

We choose a definite K, and may then write down the continuity requirements

	at $x = 0$	at $x = b$
ψ:	$A+B = C+D$	$C\exp(ik_2b)+D\exp(-ik_2b) = e^{iKl}\{A\exp(-ik_1a)+B\exp(ik_1a)\}$
$\partial\psi/\partial x$:	$ik_1(A-B) = ik_2(C-D)$	$ik_2\{C\exp(ik_2b)-D\exp(-ik_2b)\} = ik_1\, e^{iKl}\{A\exp(-ik_1a)-B\exp(ik_1a)\}$.

The periodicity condition has been used to obtain certain values at $x = b$ from those at $x = -a$. These continuity requirements have the form of four homogeneous linear relations between the four coefficients A, B, C, and D. For a non-trivial result a determinant must vanish:

$$\begin{vmatrix} 1 & 1 & 1 & 1 \\ ik_1 & -ik_1 & ik_2 & -ik_2 \\ e^{iKl}\exp(-ik_1a) & e^{iKl}\exp(ik_1a) & \exp(ik_2b) & \exp(-ik_2b) \\ ik_1\, e^{iKl}\exp(-ik_1a) & -ik_1\, e^{iKl}\exp(ik_1a) & ik_2\exp(ik_2b) & -ik_2\exp(-ik_2b) \end{vmatrix} = 0.$$

After rearrangement, this condition reduces to

$$\cos Kl = \cos k_1a\cos k_2b - \frac{(k_1^2+k_2^2)}{2k_1k_2}\sin k_1a\sin k_2b.$$

Substituting for k_1 and k_2 in the right side gives a condition

$$\cos Kl = \Phi(E).$$

A sketch of the function Φ is given in the problems (p. 56).

For any choice of K, the condition yields a number (actually an *infinite* number) of roots E. Fix attention on *one* of these roots, and allow K to vary continuously; the root will also vary continuously, and its possible values will therefore fill a continuous **band.** This will be true for each of the infinite number of roots, and we therefore expect the energy spectrum to consist of an infinite number of bands. This expectation can be fully justified (problem 4 below).

PROBLEMS

1. Evaluate the determinant on p. 55.

2. *The case* $0 < E < V$. Show that we may take

$$\psi(x) = \begin{cases} F\,e^{\lambda x} + G\,e^{-\lambda x} & (-a < x < 0), \\ C\exp(ik_2 x) + D\exp(-ik_2 x) & (0 < x < b), \end{cases}$$

and explain why *both* $\exp(\lambda x)$ and $\exp(-\lambda x)$ are permitted to appear. Obtain the condition

$$\cos Kl = \cosh \lambda a \cos k_2 b + \frac{(\lambda^2 - k_2^2)}{2\lambda k_2} \sinh \lambda a \sin k_2 b.$$

3. Show that substituting for λ and k_2 in the above condition leads to

$$\cos Kl = \Phi(E)$$

with the same function Φ as before. (There is thus really no distinction between the cases $0 < E < V$ and $V < E$.)

4. The graph of $\Phi(E)$ for a particular choice of V, a, and b is sketched in the figure (not to scale). Use this graph to locate the lowest energy bands. (*Hint:* $|\Phi(E)| \leqslant 1$.) (In one dimension, the bands do not overlap and the energy spectrum contains **forbidden zones.**)

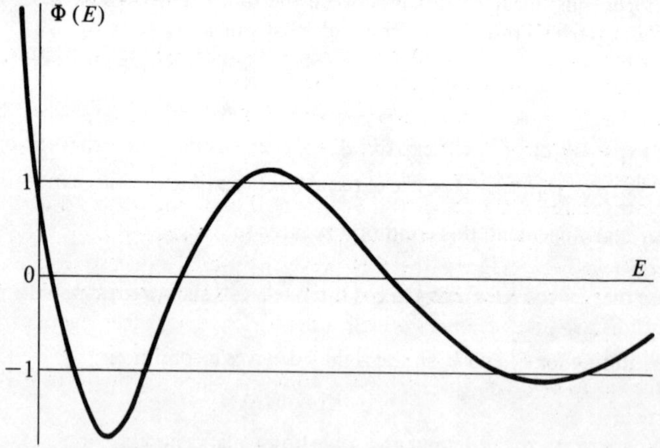

Virtual energy levels

Let us now consider a problem of a type often met in situations of *instability* or *decay*. A typical form for the potential is

$$V(x) = \begin{cases} 0 & (0 \leqslant x \leqslant a), \\ V\,(\text{constant}) & (a < x < b), \\ 0 & (b \leqslant x); \end{cases}$$

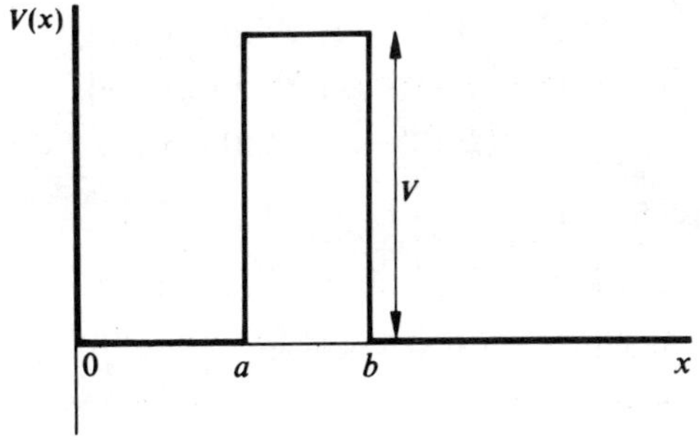

we shall assume that there is an impermeable barrier at $x = 0$, so that negative values of x are not relevant. The corresponding boundary condition is of a kind not already met; it is $\psi(0) = 0$ (problem 1 below).

The point of considering such a model becomes clear if we think of what will happen if in the region $a < x < b$, V becomes *very large and positive.* An *almost impenetrable* barrier will then have been set up. In the limit of $V = \infty$, the problem breaks up into two components, since the barrier is then completely impenetrable:

(1) the particle in the box $0 \leqslant x \leqslant a$, with a discrete energy spectrum (p. 47); for $x \geqslant a$, $\psi(x) = 0$ (problem 2 below);

(2) The particle in the infinite region $b \leqslant x$, with a continuous energy spectrum; for $0 \leqslant x \leqslant b$, $\psi(x) = 0$.

The energy spectra *overlap.*

Now imagine what happens if V is finite, though still large. As we shall see soon, the eigenvalue spectrum is now simply continuous, and the discrete energy levels are no longer there. However, a residual influence remains, in the sense that for energies near the old discrete levels the stationary solutions display 'discrete-like' features which are not found at other energies. These solutions are characterized by $\psi(x)$ taking *large* values in $0 \leqslant x \leqslant a$ compared to the values in $b \leqslant x$, i.e. they are the solutions which are closest to those for a particle in a box.

Now we turn to the details, and show how to carry out the complete solution for the stationary states. The Schrödinger equation for different regions takes the forms

$$\left.\begin{array}{l}\text{for } 0<x<a\\ \text{and } b<x\end{array}\right\} \quad -\frac{\hbar^2}{2m}\frac{\partial^2\psi}{\partial x^2}=E\psi,$$

$$\text{for } a<x<b \quad -\frac{\hbar^2}{2m}\frac{\partial^2\psi}{\partial x^2}+V\psi=E\psi.$$

The wavefunction must be written in three pieces, and (since we have in mind the case of large V) we shall take $0<E<V$:

$$\psi(x)=\begin{cases} A\,e^{ikx}+B\,e^{-ikx} & (0\leqslant x\leqslant a),\\ C\,e^{\lambda x}+D\,e^{-\lambda x} & (a\leqslant x\leqslant b),\\ E\,e^{ikx}+F\,e^{-ikx} & (b\leqslant x),\end{cases}$$

with k and λ defined by

$$\hbar^2k^2=2mE \quad \text{and} \quad \hbar^2\lambda^2=2m(V-E).$$

The boundary requirements in this case are

	ψ	$\partial\psi/\partial x$
$x=0$	$0=A+B$	—
$x=a$	$A\,e^{ika}+B\,e^{-ika}$	$ik(A\,e^{ika}-B\,e^{-ika})$
	$=C\,e^{\lambda a}+D\,e^{-\lambda a}$	$=\lambda(C\,e^{\lambda a}-D\,e^{-\lambda a})$
$x=b$	$C\,e^{\lambda b}+D\,e^{-\lambda b}$	$\lambda(C\,e^{\lambda b}-D\,e^{-\lambda b})$
	$=E\,e^{ikb}+F\,e^{-ikb}$	$=ik(E\,e^{ikb}-F\,e^{-ikb})$.

Here we have five homogeneous equations in six unknowns, and it is therefore to be expected that these have a solution for each E in the range $0<E<V$: the energy spectrum is *continuous*. Incidentally, without loss of generality we may take C, D real and A, B, E, F pure imaginary; then $\psi(x)$ is real everywhere.

There is nothing particularly remarkable about the solution of this problem. Its physical interest lies in what happens when λ is large. (This can come about when V is large and E is moderate, i.e., when a particle of moderate energy is restricted by a high potential barrier.) To see the implications, solve the three equations *not* containing E and F for $C\exp(\lambda a)$ and $D\exp(-\lambda a)$, obtaining $B=-A$ and

$$C\,e^{\lambda a}=\tfrac{1}{2}A\left\{\left(1+\frac{ik}{\lambda}\right)e^{ika}-\left(1-\frac{ik}{\lambda}\right)e^{-ika}\right\},$$

$$D\,e^{-\lambda a}=\tfrac{1}{2}A\left\{\left(1-\frac{ik}{\lambda}\right)e^{ika}-\left(1+\frac{ik}{\lambda}\right)e^{-ika}\right\}.$$

Now for *most* energies (see below), $C \exp(\lambda a)$ and $D \exp(-\lambda a)$ will be comparable in magnitude, and we have

$$\psi(a) = C\,e^{\lambda a} + D\,e^{-\lambda a}$$

$$\psi(b) = C\,e^{\lambda b} + D\,e^{-\lambda b}$$

$$= C\,e^{\lambda a}\,.\,e^{\lambda(b-a)} + D\,e^{-\lambda a}\,.\,e^{-\lambda(b-a)}.$$

Thus, even though C and D contribute to much the same extent to $\psi(a)$, in $\psi(b)$ the term in C strongly dominates the term in D, by the factor $\exp\{2\lambda(b-a)\}$, which may be very large. Thus roughly, for *most* energies,

$$\psi(b) \sim \psi(a)\,e^{\lambda(b-a)};$$

the wavefunction to the right of the barrier is much *greater* than that to the left (see figure, case (a)).

The important exceptions occur for those energies for which C is zero or very nearly zero. In such an event the term which normally dominates is absent, and we have

$$\psi(b) \sim \psi(a)\,e^{-\lambda(b-a)};$$

the wavefunction to the right is much *smaller* than that to the left. The energies for which this occurs are in the immediate neighbourhoods of those for which $C = 0$, i.e.

$$\left(1+\frac{ik}{\lambda}\right)e^{ika} - \left(1-\frac{ik}{\lambda}\right)e^{-ika} = 0,$$

or

$$\lambda + k\cot ka = 0;$$

a typical wavefunction is sketched (case (b)). These special energies are the **virtual energy levels.**

The relevance to *decay* may be seen by considering a wavefunction like the one sketched at (c). This differs little from the virtual state (b), and therefore is *nearly stationary.* It represents a state for which (to begin with) the particle has *zero* probability of being to the right of the barrier. Since the state is not quite stationary, the probability of the particle's being found to the right of the barrier is expected to grow as time passes; however, the rate of growth may be slow (it would be zero if the state were *exactly* stationary). Such a state provides a model of a decay process where a particle, initially confined behind a high barrier, ultimately **'tunnels'** through the barrier, thus becoming free.

The **lifetime** of a virtual energy level is a measure of how long we may expect to wait before finding that tunnelling has taken place; it is closely related to the ratio of the amplitude of $\psi(x)$ inside the box to that outside the

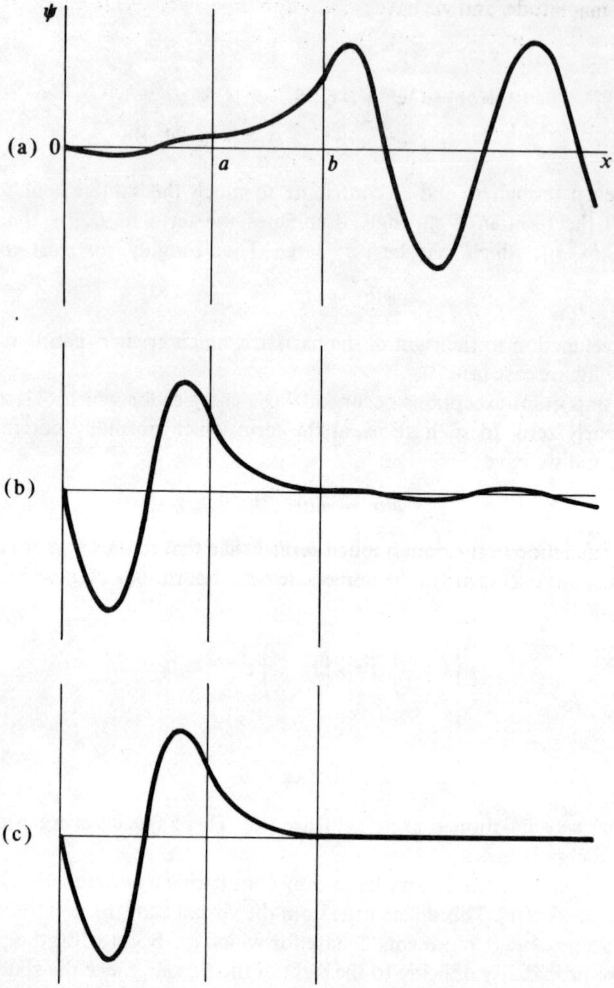

box, and this is roughly exp $\{\lambda(b-a)\}$ in the example we have considered. Since this ratio depends *exponentially* on the width $(b-a)$ and the height (through λ) of the barrier, its possible values may cover a very wide range, even for quite moderate changes in V or $(b-a)$. Experimentally, it is found that the lifetimes of 'real-life' unstable systems (such as a radioactive nucleus, an unstable elementary particle, or an atom in an excited electronic state) may have very different values, from millions of years to times so short that they cannot be measured precisely.

A word of caution: as time passes, the functions sketched in the figure do not remain real. The stationary state functions (a) and (b) keep their shape, being merely multiplied by a time-dependent phase factor $e^{-i\omega t}$ for the appropriate ω. The nonstationary state represented by the function (c) goes complex in a more complicated way.

PROBLEMS

1. In problem 1, p. 51, consider the effect of allowing V (and hence λ) to go to infinity. Show that $F \to 0$, and that the boundary condition at an *impenetrable barrier* is $\psi = 0$.

2. **A particle in a box:** A particle of mass m is confined to the region $0 \leqslant x \leqslant a$ on the x-axis; at $x = 0$ and $x = a$ there are impenetrable barriers. The requirements on the wavefunction for any stationary state are

$$-\frac{\hbar^2}{2m}\frac{\partial^2\psi}{\partial x^2} = E\psi, \quad \psi(0) = 0, \quad \psi(a) = 0.$$

Show that the energy eigenvalues and normalized eigenfunctions form the discrete set

$$E_n = \hbar^2\pi^2n^2/2ma^2 \quad \text{and} \quad \psi_n = (2/a)^{\frac{1}{2}} \sin n\pi x/a,$$

for $n = 1, 2, 3, \ldots$.

6. Quantum particles in three dimensions

Schrödinger's equation in three dimensions

THE development of the theory of a quantum particle in three dimensions is similar to that in one dimension. We may divide all space into cubes (say) of volume δ^3, and carry out the limiting operation $\delta \to 0$; we would conclude that to represent the quantum state of the particle we need a wavefunction $\psi(x, y, z, t) \equiv \psi(\boldsymbol{r}, t)$, of cartesian coordinates x, y, z (or, equivalently, of position $\boldsymbol{r}$), as well as of the time. The physical meaning of the wavefunction is

$$\left.\begin{array}{l}\text{probability at time } t \text{ of finding the particle in the}\\ \text{small volume } \mathrm{d}\tau \text{ centred on the position } \boldsymbol{r}\end{array}\right\} = |\psi(\boldsymbol{r}, t)|^2 \,\mathrm{d}\tau.$$

The total probability is 1; thus we have the normalization requirement

$$\int_{\text{all space}} |\psi(\boldsymbol{r}, t)|^2 \,\mathrm{d}\tau = 1,$$

for any physically realizable state.

Again we introduce the idea of continuity by assuming that what happens in any cube depends only on what happens in the six immediately adjacent cubes. In the limit $\delta \to 0$ we are led to the Schrödinger equation

$$\begin{aligned} \mathrm{i}\hbar \frac{\partial \psi}{\partial t}(\boldsymbol{r}, t) &= -\frac{\hbar^2}{2m}\left(\frac{\partial^2\psi}{\partial x^2}+\frac{\partial^2\psi}{\partial y^2}+\frac{\partial^2\psi}{\partial z^2}\right) + V(\boldsymbol{r})\psi \\ &= -\frac{\hbar^2}{2m}\nabla^2\psi + V(\boldsymbol{r})\psi. \end{aligned}$$

(As in one dimension, potentials like V may be introduced in more complicated ways; as in one dimension, this particular form has a very wide application.)

Observables

By considering mean values in the light of the correspondence principle, exactly as for one dimension, we are able to identify the observables for the important dynamical concepts, as listed now:

	classical concept	observable	
mass	m	m	constant
position	$\boldsymbol{r} = (x, y, z)$	$\boldsymbol{r} = (x, y, z)$	multiplication operators

classical concept		observable	
velocity	$\boldsymbol{v}$	$-\dfrac{i\hbar}{m}\nabla = \left(-\dfrac{i\hbar}{m}\dfrac{\partial}{\partial x}, -\dfrac{i\hbar}{m}\dfrac{\partial}{\partial y}, -\dfrac{i\hbar}{m}\dfrac{\partial}{\partial z}\right)$	
linear momentum	$m\boldsymbol{v}$	$-i\hbar\nabla$	derivative operators
potential energy	$V(\boldsymbol{r})$	$V(\boldsymbol{r})$	multiplication operator
force	$\boldsymbol{F} = -\nabla V$	$\boldsymbol{F} = -\nabla V$	multiplication operators
kinetic energy	$\frac{1}{2}mv^2$	$-\dfrac{\hbar^2}{2m}\nabla^2$	second-order derivative operator
total energy	$\frac{1}{2}mv^2 + V$	$H = -\dfrac{\hbar^2}{2m}\nabla^2 + V$	the Hamiltonian.

PROBLEM

(This may be omitted on a first reading.)

Fill in the details of the last two sections. (Follow the pattern of Chapter 4; the mean value of an operator O is

$$\langle O\rangle_\psi = \int_{\text{all space}} \psi^*(\mathbf{r}, t) \, . \, O\psi(\mathbf{r}, t) \, d^3\tau,$$

while three-dimensional 'integration by parts' is a version of Green's theorem:

$$\int_{\text{all space}} \psi^*(\nabla\psi) \, d^3\tau = - \int_{\text{all space}} (\nabla\psi^*) \, \psi \, d^3\tau + \text{integral over surface of infinite sphere.}$$

In all applications, the surface integral is assumed to vanish. The student who completes this question may regard himself as proficient in three-dimensional integral calculus!)

The free particle in three dimensions

The observables for the three components of linear momentum $\boldsymbol{p}$ of a particle are

$$-i\hbar\partial/\partial x, \qquad -i\hbar\partial/\partial y, \quad \text{and} \quad -i\hbar\partial/\partial z.$$

It is possible to find a **simultaneous eigenfunction** $\psi(\boldsymbol{r}; \boldsymbol{p})$ of these observables, with eigenvalues p_x, p_y, p_z; such an eigenfunction will simultaneously satisfy the three equations

$$-i\hbar\frac{\partial\psi(\boldsymbol{r};\boldsymbol{p})}{\partial x} = p_x\psi(\boldsymbol{r};\boldsymbol{p}), \qquad -i\hbar\frac{\partial\psi(\boldsymbol{r};\boldsymbol{p})}{\partial y} = p_y\psi(\boldsymbol{r};\boldsymbol{p}),$$

$$-i\hbar\frac{\partial\psi(\boldsymbol{r};\boldsymbol{p})}{\partial z} = p_z\psi(\boldsymbol{r};\boldsymbol{p}).$$

The solution is immediate:

$$\psi(\boldsymbol{r};\boldsymbol{p}) = \text{constant} \times \exp\left\{\frac{\mathrm{i}}{\hbar}(p_x x + p_y y + p_z z)\right\} = \text{constant} \times \exp(\mathrm{i}\boldsymbol{p}\,.\,\boldsymbol{r}/\hbar),$$

and, given the vector $\boldsymbol{p}$, there is just one state which corresponds to $\boldsymbol{p}$, with wavefunction $\psi(\boldsymbol{r};\boldsymbol{p})$; it represents a beam of particles, all with the definite momentum vector $\boldsymbol{p}$, and therefore the definite velocity $\boldsymbol{p}/m$ and kinetic energy $p^2/2m$. We call it the eigenstate of linear momentum $\boldsymbol{p}$.

The *free* particle in three dimensions has only kinetic energy, and no potential energy. The Schrödinger equation is therefore

$$\mathrm{i}\hbar\frac{\partial\psi}{\partial t}(\boldsymbol{r},t) = -\frac{\hbar^2}{2m}\nabla^2\psi(\boldsymbol{r},t).$$

For any fixed vector $\boldsymbol{p}$, the function

$$\psi = \exp\left\{\frac{\mathrm{i}}{\hbar}(\boldsymbol{p}\,.\,\boldsymbol{r} - E_{\boldsymbol{p}}t)\right\}, \qquad \text{with } E_{\boldsymbol{p}} = p^2/2m,$$

is a solution of the Schrödinger equation, and represents a stationary state with energy $E_{\boldsymbol{p}}$. *Every eigenstate with momentum* $\boldsymbol{p}$ *is a stationary state of the free particle* with energy $E_{\boldsymbol{p}}$.

The reverse is not true. There are many linear momenta $\boldsymbol{p}$ with the *same* $E_{\boldsymbol{p}}$ (>0). The eigenvalues of the Hamiltonian are thus highly degenerate, and there are stationary states which are not eigenstates of linear momentum.

The momentum eigenstates are plane waves: at any time, surfaces of constant ψ are planes $\boldsymbol{p}\,.\,\boldsymbol{r}$ = constant, all of which are perpendicular to the direction of propagation $\boldsymbol{p}$. They are often called **de Broglie waves**, from a suggestion first made by de Broglie in 1923 that quantum particles should have waves of this type associated with them.

PROBLEMS

1. Show that $\psi = r^{-1}\sin kr \exp(-\mathrm{i}Et/\hbar)$, with $r = (x^2+y^2+z^2)^{\frac{1}{2}}$ and $E = \hbar^2k^2/2m$, is a stationary state of the free particle, but that it is not an eigenstate of linear momentum.

2. Show that $r^{-1}\sin kr$ is a linear superposition of all the momentum eigenstates $\exp(\mathrm{i}\boldsymbol{k}\,.\,\boldsymbol{r})$ with $|\boldsymbol{k}| = k$. (First show that

$$r^{-1}\sin kr = \frac{k}{4\pi}\int_0^{\pi}\sin\theta\,\mathrm{d}\theta\int_0^{2\pi}\mathrm{d}\phi\,.\,\exp(\mathrm{i}kr\cos\theta).)$$

3. Show that, if $|\boldsymbol{p}_1|^2 = |\boldsymbol{p}_2|^2 = 2mE$, the function

$$\psi = \{\exp(\mathrm{i}\boldsymbol{p}_1\,.\,\boldsymbol{r}/\hbar) - \exp(\mathrm{i}\boldsymbol{p}_2\,.\,\boldsymbol{r}/\hbar)\}\exp(-\mathrm{i}Et/\hbar)$$

represents a stationary state of the free particle which is *always zero* on the planes

$$(\boldsymbol{p}_1 - \boldsymbol{p}_2) \cdot \boldsymbol{r} = 2\pi n\hbar,$$

where n is any integer. (Note that this eigenstate is not a plane-wave solution. Describe the physical state it represents.)

The Davisson–Germer experiment

In 1923 Davisson published some work on the angular distribution of electrons reflected from a sheet of platinum; he found the curious result that at certain angles there were strong maxima in the intensity of the elastically reflected beam. Born suggested that this might be an interference phenomenon related to de Broglie's ideas on the connection of plane waves with quantum particles. This in turn led Davisson and Germer (in 1927) to scatter slow electrons from the surface of a nickel single crystal; they found that electrons were reflected preferentially along certain moderately well-defined directions. In fact, the results were closely similar to those for the scattering of X-rays, thus lending strong support to de Broglie's ideas.

We shall consider a rather idealized model of a crystal structure, namely, an array of identical 'point scatterers' situated at the vertices of a crystal lattice. A 'point scatterer' at the point $\boldsymbol{r}_0$ could be, for example, a potential well occupying a very small volume, but deep enough to produce a non-negligible scattering:

$$V(\boldsymbol{r}) = \begin{cases} V_0 & |\boldsymbol{r} - \boldsymbol{r}_0| < a, \\ 0 & \text{otherwise,} \end{cases}$$

with a very small and V_0 very large. A plane wave falling on such an array of scatterers will in general be scattered in a very complicated and rather confused way; thus the stationary states are not easy to find.

There are, however, notable exceptions to this. The wavefunction

$$\psi = \exp(\mathrm{i}\boldsymbol{p}_1 \cdot \boldsymbol{r}/\hbar) - \exp(\mathrm{i}\boldsymbol{p}_2 \cdot \boldsymbol{r}/\hbar), \qquad \text{with } p_1^2 = p_2^2 = 2mE,$$

is *zero* on the planes $(\boldsymbol{p}_1 - \boldsymbol{p}_2) \cdot \boldsymbol{r} = 2\pi n\hbar$ (n any integer). On account of the regularity of the crystal lattice it is always possible to adjust $\boldsymbol{p}_1$ and $\boldsymbol{p}_2$ so that *every vertex of the lattice lies in one of the planes*. Then the potential is non-zero only where ψ is zero, and the term $V\psi$ is zero everywhere for wavefunctions of this special kind. It follows that these wavefunctions represent stationary states for this problem.

Such a wavefunction describes an incident beam of momentum $\boldsymbol{p}_1$ being reflected into a beam of momentum *precisely* equal to $\boldsymbol{p}_2$. This is an instance of strong scattering along a special direction, and it fully explains the Davisson–Germer result and provides powerful confirmation of the validity of quantum mechanics at the atomic level.

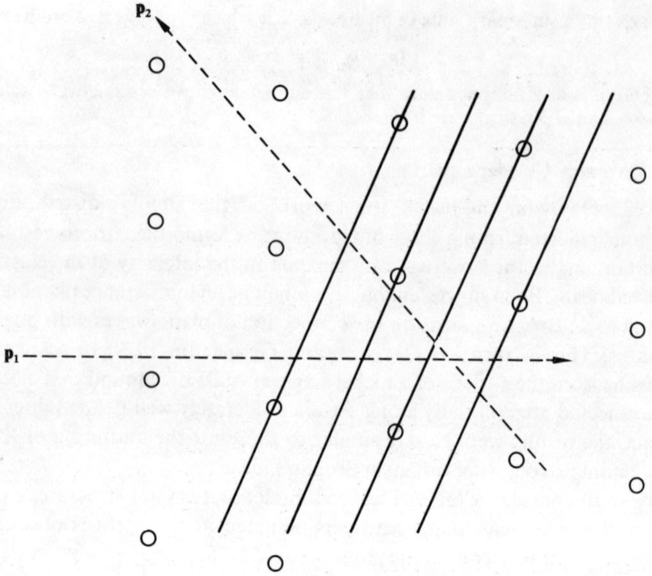

Scattering from a lattice. ○ lattice sites; ——— planes of zero ψ; - - - -→ directions of incident and reflected momenta.

It should be said that it is not at all necessary to use point scatterers to obtain results of this kind. The important feature is that the lattice has periodic properties which must be properly matched with the wavefunction periodicity embodied in the relation $(\boldsymbol{p}_1 - \boldsymbol{p}_2) \cdot \boldsymbol{r} = nh$. Ideas of this kind were first proposed in 1913 by W. L. Bragg in connection with the apparently quite different problem of the scattering of X-rays by a crystal. Similar phenomena are observed for the scattering of neutrons, helium atoms, and even hydrogen molecules.

PROBLEM

Given that $|\boldsymbol{p}_1| = |\boldsymbol{p}_2| = p$, that the planes of zero wavefunction are a distance d apart, and that the angle of incidence of the beam on the planes is ϕ, show that

$$(\boldsymbol{p}_1 - \boldsymbol{p}_2) \cdot \boldsymbol{r} = nh$$

leads to the scattering law

$$2pd \cos \phi = nh.$$

Relate this with the **Bragg scattering law** for X-rays

$$2d \cos \phi = n\lambda.$$

7. The Stern–Gerlach effect and the spin of the electron

The Stern–Gerlach experiment

IN classical physics, a particle (such as a neutral atom) with a magnetic moment $\boldsymbol{\mu}$ will in the presence of an inhomogeneous magnetic field $\boldsymbol{B}(\boldsymbol{r})$ experience a *force* $\boldsymbol{\mu} . \nabla \boldsymbol{B}$; it follows that a particle passing through such a field will generally be deflected, the amount and direction of the deflection depending on the orientation of $\boldsymbol{\mu}$.

In 1921, O. Stern and W. Gerlach used this fact to design an experiment to measure the magnetic moment of an atom of silver. A narrow collimated beam of silver atoms was passed through a field and then allowed to strike and adhere to a glass plate; the field itself was made highly non-uniform by careful design of the polepieces. The classical expectation was that the beam should be broadened in the presence of the field, and that the amount of broadening should give an estimate of μ. This was not observed; the beam was found to be *split* into two distinct components.

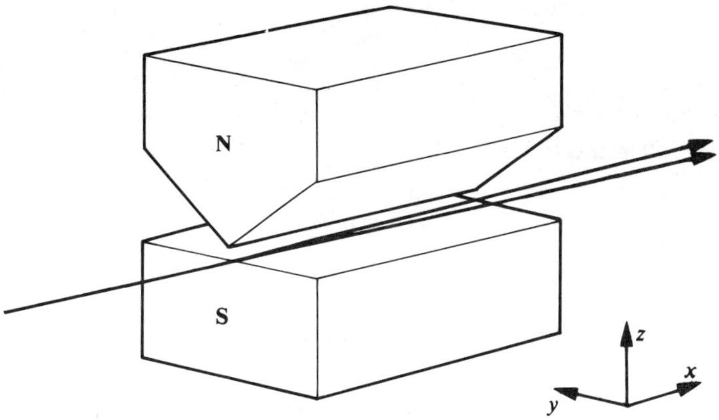

This is incomprehensible to classical physics. By contrast, quantum mechanics can provide a satisfying description of the effect, and we shall see that it even predicts a remarkable relation between magnetic moment and angular momentum which has been fully confirmed by experiment.

On the basis of the modern theory of atomic structure we now understand that the magnetic moment of the silver atom (in the ground state) arises from an *intrinsic* polarization of the single valence electron; thus much of the work of this chapter may be regarded as applying interchangeably to silver atoms or to electrons. However, it must be mentioned that the Stern–Gerlach experiment in its original form cannot be carried out with electron beams; the small effects sought are masked by the very large deflection of anything other than neutral or massive particles.

Electron polarization: the nature of the problem

In this chapter we shall adopt a deductive approach, i.e., we shall assume the rules of quantum mechanics along with some reasonable ideas about the isotropy of space, and use the Stern–Gerlach result to lead us to the correct quantum description of electron polarization. This approach gives a useful illustration of how quantum mechanics may be used in practice.

To begin with, we shall consider the observables for the magnetic moment of the electron separately from those for position, momentum, etc., just as we did for photons in Chapter 1. The Stern–Gerlach experiment is in fact an attempt to measure the component of magnetic moment in a specified direction (indicated by a unit vector $\boldsymbol{l}$, say). The quantum interpretation of the splitting of the beam into two is then that a measurement of the $\boldsymbol{l}$-component of the magnetic moment will result in one or other of only two possible values, μ or $-\mu$. This fact implies that the relevant observables are 2×2 matrices; let us write $\mu\sigma_{\boldsymbol{l}}$ for the observable for a measurement of the $\boldsymbol{l}$-component, where μ is the magnitude of the magnetic moment and $\sigma_{\boldsymbol{l}}$ is a 2×2 matrix with eigenvalues 1 and -1. Setting up the quantum mechanics of electron polarization means finding $\sigma_{\boldsymbol{l}}$ for each unit vector $\boldsymbol{l}$.

The Pauli matrices

We begin by choosing a right-handed set of x-, y-, and z-axes, and arbitrarily take for $\sigma_{\boldsymbol{l}}$, with $\boldsymbol{l} = (0, 0, 1)$, the observable

$$\sigma_z = \begin{pmatrix} 1 & 0 \\ 0 & -1 \end{pmatrix} \left\{ \begin{array}{lcc} \text{eigenvalues} & 1 & -1 \\ \text{eigenvectors} & \begin{pmatrix} 1 \\ 0 \end{pmatrix} & \begin{pmatrix} 0 \\ 1 \end{pmatrix} \\ \text{(representing, say,} & \mathrm{Q}_z & \mathrm{Q}_{-z}). \end{array} \right.$$

The arbitrariness will be discussed later.

Now consider $\boldsymbol{l}$ in the xy-plane. Suppose the electron is in the state Q_z, just defined. Measuring the $\boldsymbol{l}$-component of the magnetic moment will give either μ or $-\mu$, and we assert that because $\boldsymbol{l}$ is perpendicular to the z-axis the two results will occur *equally often* (space would not otherwise be isotropic).

Thus for the state Q_z, the mean values of σ_l is zero; similarly for Q_{-z}. Therefore for any $\boldsymbol{l}$ in the xy-plane

$$\langle\sigma_l\rangle_{Q_z} = (1 \quad 0)\sigma_l\begin{pmatrix}1\\0\end{pmatrix} = 0; \qquad \langle\sigma_l\rangle_{Q_{-z}} = (0 \quad 1)\sigma_l\begin{pmatrix}0\\1\end{pmatrix} = 0.$$

These imply that the diagonal elements of σ_l are both zero; the further fact that σ_l is Hermitian with eigenvalues ± 1 now gives

$$\sigma_l = \begin{pmatrix}0 & a_l\\ a_l^* & 0\end{pmatrix} \qquad \text{with } |a_l| = 1.$$

Arbitrarily, we take $a_l = 1$ for $\boldsymbol{l} = (1, 0, 0)$, i.e.

$$\sigma_x = \begin{pmatrix}0 & 1\\ 1 & 0\end{pmatrix}\begin{cases}\text{eigenvalues} & 1 & -1\\ \text{eigenvectors} & \frac{1}{\sqrt{2}}\begin{pmatrix}1\\1\end{pmatrix} & \frac{1}{\sqrt{2}}\begin{pmatrix}1\\-1\end{pmatrix}\\ \text{(representing, say} & Q_x & Q_{-x}).\end{cases}$$

The obvious next step is to find σ_y (i.e. σ_l for $\boldsymbol{l} = (0, 1, 0)$). Again because space is isotropic, $\langle\sigma_y\rangle = 0$ for each of the states Q_z, Q_{-z}, Q_x, Q_{-x}, and it is simple to verify that these four conditions lead to

$$\sigma_y = \begin{pmatrix}0 & a_y\\ a_y^* & 0\end{pmatrix} \qquad \text{with } a_y = +\mathrm{i} \text{ or } -\mathrm{i}.$$

Arbitrarily, we choose $a_y = -\mathrm{i}$, and thus

$$\sigma_y = \begin{pmatrix}0 & -\mathrm{i}\\ \mathrm{i} & 0\end{pmatrix}\begin{cases}\text{eigenvalues} & 1 & -1\\ \text{eigenvectors} & \frac{1}{\sqrt{2}}\begin{pmatrix}1\\ \mathrm{i}\end{pmatrix} & \frac{1}{\sqrt{2}}\begin{pmatrix}1\\-\mathrm{i}\end{pmatrix}\\ \text{(representing} & Q_y & Q_{-y}).\end{cases}$$

The arbitrary choices in this work are inevitable, since they merely mirror the arbitrariness in setting up rectangular axes: (1) choose a z-axis (thus fixing the xy-plane); (2) choose an x-axis in this plane (thus fixing the line of the y-axis); (3) choose the sense of the y-axis. Apart from this, there is no room for manoeuvre, and the quantum representation of electron magnetic moment is essentially unique.

The matrices σ_x, σ_y, and σ_z are the **Pauli matrices;** they have been universally adopted as observables for electron magnetic moment in right-handed axes. Pauli in 1925 was one of the first to recognize the twofold nature of electron polarization as an explanation of some puzzling features of atomic spectra in the presence of a magnetic field.

PROBLEMS

1. Carry out the algebra of this section in detail.

2. Show that
   (a) $\sigma_x^2 = \sigma_y^2 = \sigma_z^2 = 1$,
   (b) $\sigma_x\sigma_y = \mathrm{i}\sigma_z$, and two similar relations,
   (c) $\sigma_x\sigma_y\sigma_z = \mathrm{i}$.

The observable σ_l for general l

Now imagine a very large number N of electrons, every one in the same quantum state Q. Since magnetic moment is *additive*, the magnetic moment $\boldsymbol{M}$ of the entire collection will be an ordinary classical vector whose components are determined by mean values for the state Q (this is the correspondence principle again):

$$\boldsymbol{M} = N\mu(\langle\sigma_x\rangle_{\mathrm{Q}}, \langle\sigma_y\rangle_{\mathrm{Q}}, \langle\sigma_z\rangle_{\mathrm{Q}}).$$

The component of $\boldsymbol{M}$ in the direction of the unit vector $\boldsymbol{l}$ is $\boldsymbol{l}\,.\,\boldsymbol{M}$, and this is related (by the correspondence principle) to the mean value of the (as yet unknown) observable σ_l; in fact,

$$\boldsymbol{l}\,.\,\boldsymbol{M} = N\mu\langle\sigma_l\rangle_{\mathrm{Q}}.$$

Consequently,

$$\langle\sigma_l\rangle_{\mathrm{Q}} = l_x\langle\sigma_x\rangle_{\mathrm{Q}} + l_y\langle\sigma_y\rangle_{\mathrm{Q}} + l_z\langle\sigma_z\rangle_{\mathrm{Q}}.$$

As this is true for any state Q, we may immediately identify the observable

$$\begin{aligned}\sigma_l &= l_x\sigma_x + l_y\sigma_y + l_z\sigma_z \\ &= \boldsymbol{l}\,.\,\boldsymbol{\sigma}.\end{aligned}$$

This last expression is written according to a natural convention in which the Pauli matrices are the three components of a 'matrix vector' $\boldsymbol{\sigma}$. Indeed, one may expect that, since it represents a magnetic moment, $\mu\boldsymbol{\sigma}$ must have important three-dimensional vector properties.

This completes the representation of the electron magnetic moment by quantum observables. It remains to examine the consequences of this representation.

PROBLEMS

1. Verify that the eigenvalues of $\boldsymbol{l}\,.\,\boldsymbol{\sigma}$ are 1 and -1.

2. Show that
   (a) $(\boldsymbol{l}\,.\,\boldsymbol{\sigma})(\boldsymbol{m}\,.\,\boldsymbol{\sigma}) = \boldsymbol{l}\,.\,\boldsymbol{m} + \mathrm{i}(\boldsymbol{l}\wedge\boldsymbol{m})\,.\,\boldsymbol{\sigma}$,
   (b) $(\boldsymbol{l}\,.\,\boldsymbol{\sigma})(\boldsymbol{m}\,.\,\boldsymbol{\sigma})(\boldsymbol{n}\,.\,\boldsymbol{\sigma}) = \mathrm{i}(\boldsymbol{l}\wedge\boldsymbol{m})\,.\,\boldsymbol{n} + \{(\boldsymbol{l}\,.\,\boldsymbol{m})\boldsymbol{n} - (\boldsymbol{l}\,.\,\boldsymbol{n})\boldsymbol{m} + (\boldsymbol{m}\,.\,\boldsymbol{n})\boldsymbol{l}\}\,.\,\boldsymbol{\sigma}$,
   and obtain the results of problem 2 (above) as special cases.

3. Show that *any* Hermitian 2×2 matrix with eigenvalues $+1$ and -1 may be written in the form $a\sigma_x + b\sigma_y + c\sigma_z$, where $a^2+b^2+c^2 = 1$ and a, b, and c are all real. (The point is that there are *just enough* observables to provide a σ_l for each $\boldsymbol{l}$; this is why the apparently meagre fact that the Stern–Gerlach experiment gives two beams leads so definitely to the Pauli matrices.)

Electron spin

It is a remarkable fact that, as a consequence of the two-state nature of its magnetic moment, the electron possesses an *intrinsic angular momentum* (see p. 47).

The energy of interaction of a magnetic moment $\boldsymbol{\mu}$ in a uniform magnetic field $\boldsymbol{B}$ is $-\boldsymbol{\mu}\,.\,\boldsymbol{B}$. Thus the Hamiltonian for the electron magnetic moment in such a field is $-\mu\boldsymbol{B}\,.\,\boldsymbol{\sigma}$, leading to the equation of motion for the state vector $\chi(t)$,

$$\mathrm{i}\hbar\dot{\chi} = -\mu\boldsymbol{B}\,.\,\boldsymbol{\sigma}\chi.$$

The mean values $s_x = \langle\sigma_x\rangle$, etc. are of particular interest. Their equations of motion may be combined into a single vector equation for $\boldsymbol{s} = (s_x, s_y, s_z)$,

$$\mathrm{i}\hbar\dot{\boldsymbol{s}} = \langle[\boldsymbol{\sigma}, -\mu\boldsymbol{B}\,.\,\boldsymbol{\sigma}]\rangle$$

(see p. 33). After rearrangement, we obtain

$$\frac{\mathrm{d}}{\mathrm{d}t}(\tfrac{1}{2}\hbar\boldsymbol{s}) = -\mu\boldsymbol{B}\wedge\boldsymbol{s}$$

(problem 1, below). Now the right side of this equation is precisely the *torque* exerted by the field $\boldsymbol{B}$ on the mean magnetic moment $\mu\boldsymbol{s}$; furthermore, torque is the rate of change of the mean angular momentum. By inspection therefore, the mean angular momentum of the electron is $\tfrac{1}{2}\hbar\boldsymbol{s}$.

In the spirit of the correspondence principle, we are using the fact that angular momentum and torque are both naturally additive. In this same spirit, we now call the three observables $(\tfrac{1}{2}\hbar\sigma_x, \tfrac{1}{2}\hbar\sigma_y, \tfrac{1}{2}\hbar\sigma_z)$ the components of **electron spin;** each has eigenvalues $\pm\tfrac{1}{2}\hbar$. Planck's constant once again appears as a natural unit for angular momentum.

PROBLEMS

1. Show that for any vector $\boldsymbol{a}$

$$[\boldsymbol{\sigma}, \boldsymbol{a}\,.\,\boldsymbol{\sigma}] = 2\mathrm{i}\boldsymbol{a}\wedge\boldsymbol{\sigma}.$$

(This is a set of *three* equations, the first being

$$[\sigma_x, a_x\sigma_x + a_y\sigma_y + a_z\sigma_z] = 2\mathrm{i}(a_y\sigma_z - a_z\sigma_y).)$$

2. Show that, when $\boldsymbol{B} = (0, 0, B)$, the equation of motion for χ has the solution

$$\chi(t) = \begin{pmatrix} \chi_+ \exp(-\tfrac{1}{2}\mathrm{i}\Omega t) \\ \chi_- \exp(\tfrac{1}{2}\mathrm{i}\Omega t) \end{pmatrix},$$

where $\hbar\Omega = -2\mu B$, and the complex constants $\chi_\pm$ satisfy $|\chi_+|^2 + |\chi_-|^2 = 1$.

3. Show that, when $\boldsymbol{B} = (0, 0, B)$, the equation of motion for $\boldsymbol{s}$ has the solution

$$\boldsymbol{s} = (A \cos \Omega t - B \sin \Omega t,\ A \sin \Omega t + B \cos \Omega t,\ C)$$

where $\hbar\Omega = -2\mu B$ and A, B, and C are real constants. (In this motion, the angle θ between $\boldsymbol{s}$ and $\boldsymbol{B}$ remains constant, while $\boldsymbol{s}$ 'rotates' about $\boldsymbol{B}$ with angular velocity Ω. This **precession** is very characteristic of the behaviour of an angular momentum. Moreover, an electron in a constant magnetic field B is very susceptible to incident radiation of angular frequency $2\mu B/\hbar$; this *electron spin resonance* may be used as the basis of a kind of magnetometer on the atomic scale.)

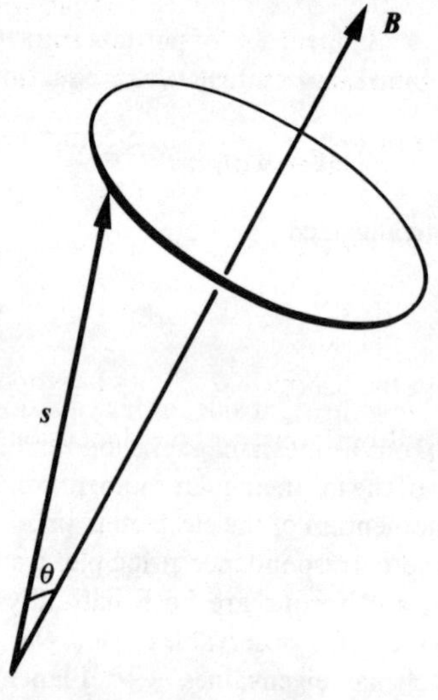

The electron does not spin!

The electron is a charged particle with an angular momentum, a magnetic moment, and a non-zero mass. It is therefore tempting to adopt a mental picture of a small spinning object with moment of inertia and charge distribution chosen to give the correct angular momentum and magnetic moment. However, this is entirely wrong.

The reason is that there is nothing in the theory corresponding to an *angular velocity* for the spin. Indeed it can be shown on very general grounds that any quantum-mechanical angular momentum which does have an angular velocity associated with it must have eigenvalues which are integer multiples of $\hbar$. The electron spin, with eigenvalues $\pm\frac{1}{2}\hbar$, cannot arise as the angular momentum of an object which is actually rotating, and it is wise to reject such intuitive pictures as misleading. The choice of the word 'spin' is a historical accident with perhaps unfortunate overtones.

The wavefunction for an electron

An electron is a quantum particle, since it has a position and a momentum. On account of its spin, it is somewhat more complicated than the quantum particle of Chapter 4, and we now consider the form that its wavefunction must take.

We start with the remark that it is possible simultaneously to measure the position of the electron and its z-component of spin (strictly, this is an assumption which has been well corroborated by experiment). The typical possible outcome of such a simultaneous measurement is then

$$\left.\begin{array}{l}\text{the position } \boldsymbol{r} \text{ of}\\ \text{the electron}\end{array}\right\}\text{taken with}\left\{\begin{array}{l}\text{the } z\text{-component } (\tfrac{1}{2}\hbar \text{ or } -\tfrac{1}{2}\hbar)\\ \text{of the spin.}\end{array}\right.$$

Corresponding to this, the general state of the electron is described by a wavefunction $\psi(\boldsymbol{r}, \sigma)$, with the interpretation

$$|\psi(\boldsymbol{r}, \sigma)|^2\,\mathrm{d}^3\boldsymbol{r} \equiv \begin{array}{l}\text{the probability that the electron is found in}\\ \text{the volume } \mathrm{d}^3\boldsymbol{r}\text{, with a } z\text{-component of spin } \tfrac{1}{2}\hbar\sigma.\end{array}$$

(Here $\boldsymbol{r}$ ranges over all positions in three dimensions, while σ has just two possible values, $+1$ and -1.)

If desired, the function $\psi(\boldsymbol{r}, \sigma)$ may be regarded as a *pair* of functions $\psi(\boldsymbol{r}, 1)$ and $\psi(\boldsymbol{r}, -1)$; these are then conventionally taken together to form a two-component column-vector function of position:

$$\psi(\boldsymbol{r}) \equiv \begin{pmatrix}\psi(\boldsymbol{r}, 1)\\ \psi(\boldsymbol{r}, -1)\end{pmatrix}.$$

The observables $\boldsymbol{r}$, $-\mathrm{i}\hbar\nabla$, and $\boldsymbol{\sigma}$ are then to be interpreted as follows:

$$x\psi(\boldsymbol{r}) = \begin{pmatrix}x\psi(\boldsymbol{r}, 1)\\ x\psi(\boldsymbol{r}, -1)\end{pmatrix}, \qquad -\mathrm{i}\hbar\frac{\partial\psi(\boldsymbol{r})}{\partial x} = \begin{pmatrix}-\mathrm{i}\hbar\partial\psi(\boldsymbol{r}, 1)/\partial x\\ -\mathrm{i}\hbar\partial\psi(\boldsymbol{r}, -1)/\partial x\end{pmatrix},$$

and

$$\sigma_x\begin{pmatrix}\psi(\boldsymbol{r}, 1)\\ \psi(\boldsymbol{r}, -1)\end{pmatrix} = \begin{pmatrix}0 & 1\\ 1 & 0\end{pmatrix}\begin{pmatrix}\psi(\boldsymbol{r}, 1)\\ \psi(\boldsymbol{r}, -1)\end{pmatrix} = \begin{pmatrix}\psi(\boldsymbol{r}, -1)\\ \psi(\boldsymbol{r}, 1)\end{pmatrix},$$

and similarly for the y- and z-components. Thus $\boldsymbol{r}$ and $-\mathrm{i}\hbar\nabla$ leave the argument σ unaffected, while the Pauli matrices leave the arguments $\boldsymbol{r}$ unaffected.

The Schrödinger equation for a free electron still has the familiar form

$$\mathrm{i}\hbar\frac{\partial\psi}{\partial t} = -\frac{\hbar^2}{2m}\nabla^2\psi.$$

A plane-wave stationary state is represented by $u\exp(\mathrm{i}\boldsymbol{k}\,.\,\boldsymbol{r}-\mathrm{i}\omega t)$, where u is a column vector with two constant components. If $\boldsymbol{l}$ is such that $\boldsymbol{l}\,.\,\boldsymbol{\sigma}u = u$, then a simultaneous measurement of momentum and the $\boldsymbol{l}$-component of spin will *certainly* yield $\hbar\mathbf{k}$ and $+\frac{1}{2}\hbar$ respectively.

PROBLEMS

1. Show that a simultaneous measurement of momentum and y-component of spin in the state $\begin{pmatrix} \exp(\mathrm{i}\boldsymbol{k}\,.\,\boldsymbol{r}) \\ -\mathrm{i}\exp(\mathrm{i}\boldsymbol{k}\,.\,\boldsymbol{r}) \end{pmatrix}$ will *certainly* yield $\hbar\mathbf{k}$ and $-\frac{1}{2}\hbar$.

2. Show that for the state $\begin{pmatrix} \exp(\mathrm{i}\boldsymbol{k}\,.\,\boldsymbol{r}) \\ \exp(-\mathrm{i}\boldsymbol{k}\,.\,\boldsymbol{r}) \end{pmatrix}$ a measurement of momentum will give $\hbar\mathbf{k}$ or $-\hbar\mathbf{k}$ with equal probability. (Evaluate the mean value $\psi^+(\boldsymbol{r})(-\mathrm{i}\hbar\nabla)\psi(\boldsymbol{r})$.)

3. Show that for the state of problem 2, a simultaneous measurement of momentum and z-component of spin will *never* yield $+\hbar\mathbf{k}$ with $-\frac{1}{2}\hbar$. Note that

$$\begin{pmatrix} \exp(\mathrm{i}\boldsymbol{k}\,.\,\boldsymbol{r}) \\ \exp(-\mathrm{i}\boldsymbol{k}\,.\,\boldsymbol{r}) \end{pmatrix} = \begin{pmatrix} 1 \\ 0 \end{pmatrix}\exp(\mathrm{i}\boldsymbol{k}\,.\,\boldsymbol{r}) + \begin{pmatrix} 0 \\ 1 \end{pmatrix}\exp(-\mathrm{i}\boldsymbol{k}\,.\,\boldsymbol{r})$$

and interpret the terms on the right. This is an example of **correlation**. For such a state, measuring momentum will tell us what measuring spin would give, *without our having to do it.* A practical example of this is the Stern–Gerlach experiment, where the spin of an electron is correlated with its direction of emergence from the apparatus; see below.

The Stern–Gerlach effect: the Schrödinger equation

A neutral silver atom in a given inhomogeneous magnetic field $\boldsymbol{B}(\boldsymbol{r})$ has a total energy comprising a kinetic energy and an energy of its magnetic moment in the field; we thus adopt as its Hamiltonian

$$H = -\frac{\hbar^2}{2m}\nabla^2 - \mu\boldsymbol{\sigma}\,.\,\boldsymbol{B}(\boldsymbol{r}).$$

We have assumed that the very complicated structure of the atom is 'felt' externally only through its magnetic moment; such an approximation is valid if $\boldsymbol{B}$ varies little over the volume of an atom, and this is certainly true of the Stern–Gerlach experiment and similar experimental situations.

To describe the magnetic field of the Stern–Gerlach experiment, we first set up axes as follows: take the x-axis along the centre of the atomic beam, and the z-axis in the direction of the field at the beam centre (see figure, p. 67). To a sufficient approximation the magnetic field vector in the vicinity of the beam is then

$$\boldsymbol{B}(x, y, z) = (0, -\alpha y, B_0 + \alpha z),$$

where α is a constant whose value gives the amount of the inhomogeneity. For this field,

$$-\mu\boldsymbol{\sigma}.\boldsymbol{B} = -\mu\begin{pmatrix} B_0+\alpha z & \mathrm{i}\alpha y \\ -\mathrm{i}\alpha y & -B_0-\alpha z \end{pmatrix},$$

leading to the Schrödinger equation for $\psi(\boldsymbol{r})$ (which we write in terms of the components $\psi_+ = \psi(\boldsymbol{r}, 1), \psi_- = \psi(\boldsymbol{r}, -1)$)

$$\mathrm{i}\hbar\frac{\partial\psi_+}{\partial t} = -\frac{\hbar^2}{2m}\nabla^2\psi_+ - \mu(B_0+\alpha z)\psi_+ - \mathrm{i}\alpha\mu y\psi_-,$$

$$\mathrm{i}\hbar\frac{\partial\psi_-}{\partial t} = -\frac{\hbar^2}{2m}\nabla^2\psi_- + \mu(B_0+\alpha z)\psi_- + \mathrm{i}\alpha\mu y\psi_+.$$

Within the approximations (1) that we use these equations near the beam (i.e. that x and y should be small) and (2) that edge effects at the ends of the pole-pieces are neglected, this provides the quantum description of the Stern–Gerlach experiment.

PROBLEM

Show that, if H is the Stern–Gerlach Hamiltonian, the mean force acting on the electron is

$$\left\langle\frac{1}{\mathrm{i}\hbar}[-\mathrm{i}\hbar\nabla, H]\right\rangle = \langle-\nabla(-\mu\boldsymbol{\sigma}.\boldsymbol{B})\rangle = \langle\alpha\mu(0, -\sigma_y, \sigma_z)\rangle.$$

For an electron in the spin state $\begin{pmatrix}1\\0\end{pmatrix}$ show that the mean force is $(0, 0, \alpha\mu)$. What is the mean force for the state $\begin{pmatrix}0\\1\end{pmatrix}$? (The mean force is the rate of change of mean momentum, $\mathrm{d}\langle\boldsymbol{p}\rangle/\mathrm{d}t$. By the work of p. 33, this is $(\mathrm{i}\hbar)^{-1}\langle[p, H]\rangle$, providing the starting point of the problem.)

The quantum explanation of the Stern–Gerlach effect

Let us first of all assume that $\alpha = 0$, i.e. the field is homogeneous. Then it is easy to show (by direct substitution) that the wavefunction

$$\psi = \begin{pmatrix}\chi_+ \exp\{\mathrm{i}(kx-\omega_+ t)\} \\ \chi_- \exp\{\mathrm{i}(kx-\omega_- t)\}\end{pmatrix},$$

with χ_+, χ_- any constants and

$$\omega_\pm = \omega \pm \tfrac{1}{2}\Omega, \qquad \hbar\omega = \frac{\hbar^2 k^2}{2m}, \qquad \hbar\Omega = -2\mu B$$

is a solution of the Schrödinger equation (but note that it is not necessarily a stationary state). It represents a beam of electrons moving along the x-direction, with a momentum $\hbar k$, in a spin state described by $\begin{pmatrix}\chi_+ \\ \chi_-\end{pmatrix}$ at $t = 0$. The presence of the external field produces a spin-precession in this state with angular velocity Ω (see problem 3, p. 72).

Let us now allow α to be non-zero. The state we have been considering is no longer an exact solution of the Schrödinger equation, but will differ from an exact solution by a *small* amount, provided α is small. A common method of taking account of this is to allow the 'constants' $\chi_\pm$ to be slowly varying functions of time; if we do this, and substitute ψ into the *full* equation with $\alpha \neq 0$, we find

$$\begin{aligned} i\hbar\dot{\chi}_+ &= \alpha\mu(-z\chi_+ - iy\chi_-\, e^{i\Omega t}), \\ i\hbar\dot{\chi}_- &= \alpha\mu(z\chi_- + iy\chi_+\, e^{-i\Omega t}). \end{aligned}$$

(We ought to be more careful; since y and z appear on the right-side of these equations, $\chi_\pm$ must be functions of y and z as well as of t. However, it turns out that, for the solution we want, the terms ignored are of *second* order in α; see problem 1, p. 77).

Next we ignore the 'precessional' terms with the factors $e^{\pm i\Omega t}$. These terms are rapidly oscillating and therefore do not affect χ_+ or χ_- very much (problem 2, p. 77). There remain

$$i\hbar\dot{\chi}_+ = -\alpha\mu z\chi_+, \qquad i\hbar\dot{\chi}_- = +\alpha\mu z\chi_-$$

with solutions

$$\chi_+ = \eta_+ \exp(i\alpha\mu z t/\hbar), \qquad \chi_- = \eta_- \exp(-i\alpha\mu z t/\hbar),$$

where η_+ and η_- are constant (at least in the approximation where α^2 and the effect of Ω are ignored).

We may now substitute back to obtain an approximate ψ:

$$\psi = \begin{pmatrix} \eta_+ \exp\{i(kx + \alpha\mu z t/\hbar - \omega_+ t)\} \\ \eta_- \exp\{i(kx - \alpha\mu z t/\hbar - \omega_- t)\} \end{pmatrix}.$$

This solution may be immediately decomposed into two 'approximately stationary' solutions,

$$\psi = \begin{pmatrix}1\\0\end{pmatrix}\eta_+ \exp\{i(\boldsymbol{k}_+ \cdot \boldsymbol{r} - \omega_+ t)\} + \begin{pmatrix}0\\1\end{pmatrix}\eta_- \exp\{i(\boldsymbol{k}_- \cdot \boldsymbol{r} - \omega_- t)\},$$

where we have written

$$\boldsymbol{k}_{\pm} = (k, 0, \pm\alpha\mu t/\hbar).$$

The first term represents a beam with momentum $\hbar\boldsymbol{k}_+$ and z-component of spin $\frac{1}{2}\hbar$, while the second represents a beam with momentum $\hbar\boldsymbol{k}_-$ and z-component of spin $-\frac{1}{2}\hbar$. Thus spin is correlated with momentum in the state represented by ψ (compare with problem 3, p. 74).

Now $\boldsymbol{k}_+$ and $\boldsymbol{k}_-$ depend on time. At $t = 0$ they are the same, but they slowly but steadily move apart as time passes. After a time t, the angle between $\boldsymbol{k}_+$ and $\boldsymbol{k}_-$ is approximately $2\alpha\mu t/\hbar k$. This change of momentum (and therefore of direction), along with the strong correlation of spin with momentum, fully explains the Stern–Gerlach result.

The role of the uniform part of the field (B_0) is interesting. It does not appear in the final result; thus the angle between the emergent beams depends only on the inhomogeneity of the field, and not on its strength. However, B_0 is essential to the result, since it is responsible for the precession about the z-axis, which in its turn allows us to ignore terms with frequency Ω. Effectively, B_0 provides a reference which determines the plane in which the beam will split.

PROBLEMS

1. By direct substitution, show that the approximate ψ obtained in this section satisfies the Schrödinger equation, provided α^2 and terms involving Ω are neglected.
2. If $\dot{\phi} = \exp(\mathrm{i}\Omega t)$, show that $\phi = (\mathrm{i}\Omega)^{-1}\exp(\mathrm{i}\Omega t) + \text{constant}$, and explain why terms in $\exp(\mathrm{i}\Omega t)$ may be dropped when Ω is large.
3. The observable for the z-component of the force on the atom is $\mu\alpha\sigma_z$ (problem, p. 75). If the atom remains between the polepieces for a time T, deduce that the change in transverse momentum (in the z-direction) is either $\mu\alpha T$ or $-\mu\alpha T$, and *nothing else*. (This kind of approach provides an alternative, and often much quicker, route to the result.)
4. The observable for the y-component of the force is $-\mu\alpha\sigma_y$. Why does this component have a negligible effect?

8. A quantum particle in a spherically symmetric potential

The Schrödinger equation with a spherically symmetric potential

IN this chapter and the next we shall consider in some detail a very important special case of the Schrödinger equation, where the potential $V(\mathbf{r})$ is spherically symmetric. In this chapter some general foundations are laid.

The potential $V(\mathbf{r})$ is **spherically symmetric** when it depends on $\mathbf{r} = (x, y, z)$ only through the combination $r = (x^2+y^2+z^2)^{\frac{1}{2}}$; thus the surfaces over which V is constant are concentric spheres $|\mathbf{r}| = \text{const}$. A potential of this kind gives rise to a **central force** (in Newtonian mechanics)

$$\mathbf{F} = -\nabla V(r) = -\frac{\mathbf{r}}{r}\frac{\partial V}{\partial r},$$

so called as its line of action always passes through the origin. The Schrödinger equation for a spinless particle of mass m_0 in the potential $V(r)$ is

$$i\hbar\frac{\partial \psi}{\partial t} = -\frac{\hbar^2}{2m_0}\nabla^2\psi + V(r)\psi.$$

It should be noted that the solutions are not necessarily spherically symmetric functions of $\mathbf{r}$, though we shall expect to find some such functions among the solutions.

There are good reasons for considering central forces. For example, a fundamentally important case is the *Coulomb potential*: imagine that an electric charge q_0 is fixed at the origin, while the quantum particle carries a charge $-q$. The electrostatic energy is then

$$V(r) = -qq_0/4\pi\varepsilon_0 r.$$

Such a potential leads us to an excellent first quantum representation of monatomic hydrogen, singly ionized helium, doubly ionized lithium, and so on. In these cases, $-q$ is the charge on the electron, while q_0 is $+Zq$, the atomic number of the nucleus being Z. (A full quantum theory of even a hydrogen atom is much more complicated than this, and we shall consider a few of the refinements later.)

An *alkali atom* (Li, Na, K, Rb, Cs, etc., with atomic numbers $Z = 3$, 11, 19, 37, 55, ...) is sometimes referred to as *hydrogen-like*, for the reason following: it consists of a massive nucleus of charge $+Zq$, surrounded by a spherically symmetric arrangement of $Z-1$ electrons which is particularly stable, for reasons which we shall not consider here. The remaining 'outer' electron moves under the influence of the rest of the system; as a first approximation this influence may be 'mimicked' by an *effective* spherically symmetric potential $V(r)$. In this way the original many-particle problem is reduced to a one-particle problem, though the potential will no longer be Coulomb. The merit of this procedure is that it is the properties of the outer electron which are almost entirely responsible for the important energy levels of an alkali atom—and indeed for its major chemical features.

The quantum theory of the Coulomb potential is important historically. Almost as soon as he had drawn up his version of quantum mechanics, Schrödinger went on in 1926 to solve exactly the problem of the energy levels of a quantum particle bound to a Coulomb potential. This was an essential first step towards a full quantum theory of the hydrogen atom, and the results were close enough to experiment to show that quantum mechanics had at that time reached a definitive stage.

Mathematical prerequisites

We may write the Schrödinger equation (for a *stationary* state) as

$$\nabla^2\psi + f(r)\psi = 0,$$

where, for the spherically symmetric potential $V(r)$,

$$f(r) = \frac{2m_0}{\hbar^2}(E - V(r)),$$

the energy eigenvalue being E.

For such an equation, it is an advantage to work in **spherical polar coordinates** r, θ, ϕ, related to x, y, and z by

$$x = r \sin\theta \cos\phi$$
$$y = r \sin\theta \sin\phi$$
$$z = r \cos\theta$$

with

$$r \geqslant 0 \quad \text{and} \quad 0 \leqslant \theta \leqslant \pi.$$

(The directions $\theta = 0, \pi$—the z-axis, that is—form the **axis** of the spherical polar coordinates, and in a sense are accidentally emphasized as a result. We shall need to return to this point below.) If we regard the wavefunction ψ as a function $\psi(r, \theta, \phi)$ of the polar coordinates, the operator ∇^2 takes the form defined by

$$\nabla^2\psi \equiv \frac{1}{r}\frac{\partial^2}{\partial r^2}(r\psi) + \frac{1}{r^2 \sin\theta}\frac{\partial}{\partial\theta}\left(\sin\theta\frac{\partial\psi}{\partial\theta}\right) + \frac{1}{r^2\sin^2\theta}\frac{\partial^2\psi}{\partial\phi^2}.$$

In polar coordinates the Schrödinger equation is **separable**, in the following sense: we search for **separated solutions** of the form

$$\psi(r, \theta, \phi) = R(r)S(\theta, \phi).$$

Now such a solution is drastically restricted in form, and it must be remembered that

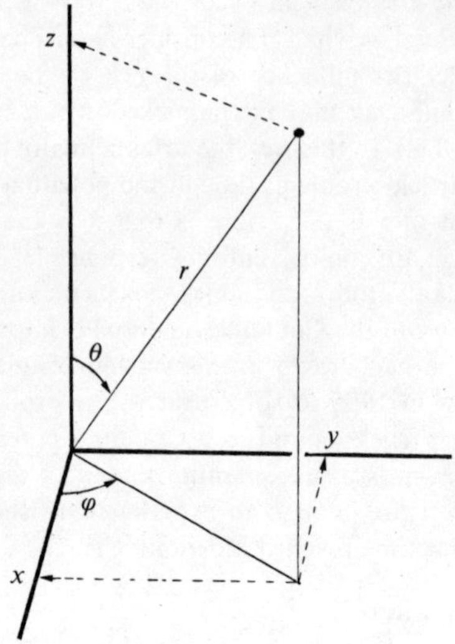

the general solution of the equation cannot be written in this way; indeed, it is not at first sight obvious that such separated solutions exist at all. They do exist, however, and there are enough of them to allow us to write *any* solution as a superposition (pp. 15, 43) of separated solutions. Thus the search for separated solutions is not as restrictive as it may seem.

Now, for a separated solution (and only for a separated solution), we may rearrange the Schrödinger equation to give

$$r^2 \frac{1}{Rr}\frac{\partial^2}{\partial r^2}(rR)+r^2 f(r) = -\frac{1}{S}\left[\frac{1}{\sin\theta}\frac{\partial}{\partial\theta}\left(\sin\theta\frac{\partial S}{\partial\theta}\right)+\frac{1}{\sin^2\theta}\frac{\partial^2 S}{\partial\phi^2}\right].$$

The left side does not depend on θ or ϕ, while the right side does not depend on r. Each side is therefore a constant α, and the functions R and S satisfy separate equations,

$$\frac{1}{r}\frac{\mathrm{d}^2}{\mathrm{d}r^2}(rR)+\left(f(r)-\frac{\alpha}{r^2}\right)R = 0 \qquad \text{(the \textbf{radial equation})},$$

$$-\frac{1}{\sin\theta}\frac{\partial}{\partial\theta}\left(\sin\theta\frac{\partial S}{\partial\theta}\right)-\frac{1}{\sin^2\theta}\frac{\partial^2 S}{\partial\phi^2} = \alpha S \qquad \text{(the \textbf{angular equation})}.$$

The value of the **separation constant** α may be different for different separated solutions.

Here we see how useful polar coordinates are. The function $f(r)$ does not appear in the angular equation. We may therefore deal with the angular equation once and for all, with results which will apply to any spherically symmetric case whatever. An important aspect of the problem is to determine the possible values of the separation constant α, which appears in the radial equation also.

The angular equation is itself separable: we may write $S(\theta, \phi) = \Theta(\theta)\Phi(\phi)$ and follow a similar procedure to obtain the two equations

$$-\frac{1}{\sin\theta}\frac{\mathrm{d}}{\mathrm{d}\theta}\left(\sin\theta\frac{\mathrm{d}\Theta}{\mathrm{d}\theta}\right)+\frac{\beta\Theta}{\sin^2\theta}=\alpha\Theta \qquad \text{(the } \Theta\text{-equation)},$$

$$\frac{\mathrm{d}^2\Phi}{\mathrm{d}\phi^2}=-\beta\Phi \qquad \text{(the } \Phi\text{-equation)},$$

where β is a further separation constant which may be different for different solutions.

We shall now see that the constants α and β cannot take arbitrary values. First, on account of the geometrical meaning of the coordinate ϕ (see figure, p. 80) the function Φ must be periodic with period 2π, in addition to satisfying the Φ-equation. Thus Φ may then be taken as any one of the functions

$$\Phi_m(\phi) = e^{im\phi}, \qquad m = \ldots -2, -1, 0, 1, \ldots,$$

corresponding to the possible values of $\beta(=m^2) = 0, 1, 4, \ldots$. (The choice of Φ is not uniquely determined if β is given: β is the same for m and for $-m$. Thus for any $\beta = m^2$, Φ may be taken to be any linear combination of Φ_m and Φ_{-m}, that is, there is twofold degeneracy for any $\beta = m^2 \neq 0$.)

The Θ-equation now becomes

$$-\frac{1}{\sin\theta}\frac{\mathrm{d}}{\mathrm{d}\theta}\left(\sin\theta\frac{\mathrm{d}\Theta}{\mathrm{d}\theta}\right)+\frac{m^2\Theta}{\sin^2\theta}=\alpha\Theta,$$

where now m must be an integer. This differential equation has so-called *regular points* at values of θ for which $\sin\theta = 0$, that is, $\theta = 0$ and π. Except in special circumstances, the solution Θ is expected to have singularities at the regular points. On physical grounds this is unacceptable: the points for which $\theta = 0$ or π form the z-axis which, as we have seen, is special only because we happen to have chosen it as the polar axis. The constant α must therefore be chosen to yield acceptable solutions; the permitted values turn out to be

$$\alpha = l(l+1) \quad \text{where} \quad l = |m|, |m|+1, |m|+2, \ldots.$$

For any such value of α there exists a solution which is a *polynomial* in $\sin\theta$ and $\cos\theta$, with no singularities anywhere. (These solutions are multiples of the **associated Legendre functions** $P_l^m(\cos\theta)$; these have a rich mathematical theory, which is beyond the scope of this book.)

The possible combinations of values for α and β may be described slightly differently; they are

$$\alpha = l(l+1) \quad \text{with} \quad l = 0, 1, 2, \ldots$$

$$\beta = m^2 \quad \text{with} \quad m = -l, -l+1, \ldots, l-1, l.$$

Thus for each choice of l, there are $2l+1$ possible choices for m. The corresponding separated solutions are **spherical harmonics** $Y_{lm}(\theta, \phi)$ (occasionally called **surface harmonics**). Those for the first few values of l are

$$l = 0 \qquad Y_{00} = (1/4\pi)^{\frac{1}{2}}$$

$$l = 1 \qquad Y_{11} = (3/8\pi)^{\frac{1}{2}} \sin\theta\, e^{i\phi}$$

$$Y_{10} = (3/4\pi)^{\frac{1}{2}} \cos\theta$$

$$Y_{1-1} = -(3/8\pi)^{\frac{1}{2}} \sin\theta\, e^{-i\phi}$$

$$l=2 \quad Y_{22}=(15/32\pi)^{\frac{1}{2}}\sin^2\theta\, e^{2i\phi}$$
$$Y_{21}=(15/8\pi)^{\frac{1}{2}}\sin\theta\cos\theta\, e^{i\phi}$$
$$Y_{20}=(5/16\pi)^{\frac{1}{2}}(3\cos^2\theta-1)$$
$$Y_{2-1}=-(15/8\pi)^{\frac{1}{2}}\sin\theta\cos\theta\, e^{-i\phi}$$
$$Y_{2-2}=(15/32\pi)^{\frac{1}{2}}\sin^2\theta\, e^{-2i\phi}.$$

(Some authors define Y_{lm} to include a further factor $(-1)^m$.)

In polar coordinates, the integral of a function $F(\boldsymbol{r})=F(r,\theta,\phi)$ over all space takes the form

$$\int_0^\infty r^2\,\mathrm{d}r\int_0^\pi \sin\theta\,\mathrm{d}\theta\int_0^{2\pi}\mathrm{d}\phi\, F(r,\theta,\phi).$$

The scalar product of two wavefunctions is thus

$$(\psi_1^*,\psi_2)=\int_0^\infty r^2\,\mathrm{d}r\int_0^\pi \sin\theta\,\mathrm{d}\theta\int_0^{2\pi}\mathrm{d}\phi\,\psi_1^*(r,\theta,\phi)\psi_2(r,\theta,\phi).$$

The separated solutions satisfy important orthogonality relations with respect to this scalar product, as may be seen from the orthonormality of the spherical harmonics:

$$\int_0^\pi \sin\theta\,\mathrm{d}\theta\int_0^{2\pi}\mathrm{d}\phi\; Y_{lm}^*(\theta,\phi)Y_{l'm'}(\theta,\phi)=\delta_{ll'}\delta_{mm'}.$$

(The orthogonality may be proved using the Θ and Φ-equations, which are simultaneously satisfied by the Ys. The unsightly numerical factors in the expressions for the Ys have been chosen to ensure that each Y_{lm} is normalized.)

It is convenient to mention at this point a slightly different presentation of the functions for $l=1$, which we shall occasionally use. Define the unit vector

$$\boldsymbol{u}(\theta,\phi)=(\sin\theta\cos\phi,\ \sin\theta\sin\phi,\ \cos\theta);$$

this is in fact the unit vector along the direction specified by θ,ϕ. A related vector is

$$\boldsymbol{Y}=\sqrt{\left(\frac{3}{4\pi}\right)}\boldsymbol{u}(\theta,\phi)=((Y_{11}-Y_{1-1})/\sqrt{2},\ (Y_{11}+Y_{1-1})/\sqrt{2},\ Y_{10}).$$

It is clear that the most general linear combination of Y_{11}, Y_{1-1}, and Y_{10} may be written $\boldsymbol{c}.\boldsymbol{Y}$, where $\boldsymbol{c}$ is an arbitrary complex *vector* in ordinary space. The three spherical harmonics

$$Y_x=\sqrt{\left(\frac{3}{4\pi}\right)}\sin\theta\cos\phi,$$
$$Y_y=\sqrt{\left(\frac{3}{4\pi}\right)}\sin\theta\sin\phi,$$
$$Y_z=\sqrt{\left(\frac{3}{4\pi}\right)}\cos\theta,$$

are just as good as the harmonics Y_{11}, Y_{10}, Y_{1-1} for the discussion of $l=1$ wavefunctions, and may be somewhat more convenient for some purposes.

Angular momentum: the significance of l

In the search for separated solutions of the Schrödinger equation we have encountered the **quantum numbers** l and m. These numbers are very convenient for indexing the different types of solution that may occur. Now it is almost always true that mathematically useful entities have an important physical significance; in this case, l and m turn out to be closely connected with *angular momentum*.

First consider l. The radial equation (p. 80) may be written

$$-\frac{\hbar^2}{2m_0}\frac{1}{r}\frac{d^2}{dr^2}(rR)+V(r)R+\frac{\hbar^2 l(l+1)}{2m_0r^2}R = ER,$$

if we remember that $f(r) = 2m_0\{E-V(r)\}/\hbar^2$, $\alpha = l(l+1)$. Let us introduce a slight change in the radial function:

$$\chi(r) = rR(r);$$

the equation to be satisfied by $\chi(r)$ is then

$$-\frac{\hbar^2}{2m_0}\frac{d^2\chi}{dr^2}+\left(V(r)+\frac{\hbar^2 l(l+1)}{2m_0r^2}\right)\chi = E\chi.$$

For any particular fixed choice of l, this equation has *precisely* the form of a Schrödinger equation for a particle of mass m_0 moving in *one* dimension in an *effective potential*

$$v_l(r) = V(r)+\frac{\hbar^2 l(l+1)}{2m_0r^2}.$$

(Of course, the coordinate r must be restricted to positive values, $r \geqslant 0$. Also, we need boundary conditions: as $r \to \infty$, χ must be bounded, while at $r = 0$, $\chi(0) = 0$; see problem 1, p. 61.)

The process of separating the variables may be regarded as a method of getting rid of the angular dependence on θ and ϕ. We cannot be completely successful in this, and a vestigial influence remains in the form of an extra term in the effective potential. To be able to give this term a name, we must turn to the analogous classical system.

Imagine therefore that a *classical* particle of mass m_0 moves in three dimensions under the influence of the potential $V(r)$. The total energy

$$E = \tfrac{1}{2}m_0\dot{\boldsymbol{r}}\,.\,\dot{\boldsymbol{r}}+V(r),$$

and the angular momentum about the origin

$$\boldsymbol{M} = m_0\boldsymbol{r}\wedge\dot{\boldsymbol{r}}$$

are both constants of the motion. The orbit followed by the particle lies entirely in the plane through O perpendicular to $\boldsymbol{M}$. Let us describe the position of the particle in this plane by the polar coordinates (r, θ) as shown. The $\boldsymbol{R}$ and $\boldsymbol{T}$ components of the velocity are then $\dot{r}$ and $r\dot{\theta}$, and we may now write

$$E = \tfrac{1}{2}m_0(\dot{r}^2 + r^2\dot{\theta}^2) + V(r)$$

and

$$|\boldsymbol{M}| = m_0 r \,.\, r\dot{\theta}.$$

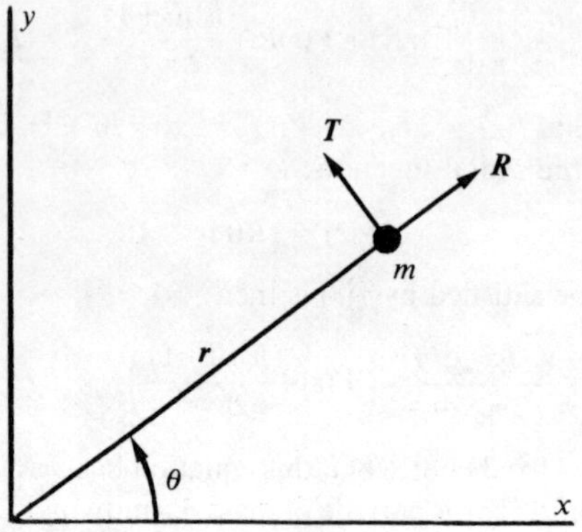

Now let us eliminate the angular motion as far as we can. Eliminating $\dot{\theta}$ between these relations gives

$$E = \tfrac{1}{2}m_0\dot{r}^2 + V(r) + \frac{\boldsymbol{M}^2}{2m_0 r^2}.$$

This is *precisely* conservation of total energy for a particle moving in *one* dimension in an *effective potential*

$$u(r) = V(r) + \frac{\boldsymbol{M}^2}{2m_0 r^2};$$

again r is restricted to $r \geqslant 0$. The vestigial influence of the angular motion comes via the (angular momentum)^2, and shows itself as the potential for the 'centrifugal' force $\boldsymbol{M}^2/mr^3$.

Comparing the effective potentials v_l and u immediately suggests (by the correspondence principle, yet again) that $\boldsymbol{M}^2$ and $\hbar^2 l(l+1)$ ought to correspond. We say that

$$L^2 \equiv (\text{total quantum angular momentum})^2 = \hbar^2 l(l+1),$$

and in this way give a meaning to the quantum number l for any state whose angular factor is $Y_{lm}(\theta, \phi)$.

It is easy to find the observable which represents L^2. The angular equation (p. 80) for the function $S = Y_{lm}$ may be written as

$$-\hbar^2\left[\frac{1}{\sin\theta}\frac{\partial}{\partial\theta}\left(\sin\theta\frac{\partial Y_{lm}}{\partial\theta}\right)+\frac{1}{\sin^2\theta}\frac{\partial^2 Y_{lm}}{\partial\phi^2}\right] = \hbar^2 l(l+1)Y_{lm}.$$

This is an eigenvalue equation which declares that Y_{lm} is an eigenfunction of the operator

$$L^2 = -\hbar^2\left[\frac{1}{\sin\theta}\frac{\partial}{\partial\theta}\left(\sin\theta\frac{\partial}{\partial\theta}\right)+\frac{1}{\sin^2\theta}\frac{\partial^2}{\partial\phi^2}\right],$$

the eigenvalue being $\hbar^2 l(l+1)$. This identifies the observable L^2 completely.

The quantum number l is often called the **orbital-angular-momentum** quantum number. It should be noted in passing that there is a convention that one often speaks of an eigenstate of 'angular momentum $\hbar l$', rather than 'angular momentum $\hbar\sqrt{(l(l+1))}$'. This is not strictly correct, but has the obvious advantage of convenience.

The significance of the quantum number *m*: the *z*-component of angular momentum

In the same way, we may find a physical significance for the integer m. Once again, we think of the *classical* situation. A particle with position $\boldsymbol{r}$ and momentum $\boldsymbol{p}$ has an angular momentum $\boldsymbol{r}\wedge\boldsymbol{p}$ about the origin. Since angular momentum is naturally additive, we may take this over into quantum mechanics, and are led to define three observables L_x, L_y, L_z, to form a vector:

$$\text{quantum angular momentum } \boldsymbol{L} = \boldsymbol{r}\wedge(-i\hbar\nabla).$$

In particular,

$$L_z = -i\hbar\left[x\frac{\partial}{\partial y}-y\frac{\partial}{\partial x}\right],$$

the observable for the component of angular momentum along the z-axis (that is, along the axis of the polar coordinates).

It is straightforward to show (problem 1) that when polar coordinates are used, L_z takes a particularly simple form

$$L_z = -i\hbar\frac{\partial}{\partial\phi}.$$

(We have met this kind of observable before, and have already noted its relevance to angular momentum (p. 47).) Now since $Y_{lm}(\theta, \phi)$ contains ϕ only in the factor $\exp im\phi$, we immediately find

$$L_z Y_{lm} \equiv -i\hbar\frac{\partial Y_{lm}}{\partial\phi} = \hbar m Y_{lm};$$

that is, Y_{lm} is an eigenfunction of L_z with eigenvalue $\hbar m$. Hence the physical significance of the quantum number m: it gives the observed value of the z-component of angular momentum. It is often called the magnetic quantum number, on account of its relevance to the Zeeman effect (chapter 13).

PROBLEMS

1. When x, y, z, r, θ, ϕ are related by $x = r \sin\theta \cos\phi$, etc. (see p. 79), show that

$$\frac{\partial x}{\partial \phi} = -y, \qquad \frac{\partial y}{\partial \phi} = x, \qquad \frac{\partial z}{\partial \phi} = 0,$$

and that
$$\frac{\partial f}{\partial \phi} = x\frac{\partial f}{\partial y} - y\frac{\partial f}{\partial x},$$

f being a function of x, y, and z.

2. From $\boldsymbol{L} = -i\hbar \boldsymbol{r} \wedge \nabla$, write down expressions for L_x and L_y, in terms of x, y, and z. Show that if f is a function of x, y, and z,

$$L_x L_y f - L_y L_x f = i\hbar L_z f.$$

(It follows that $[L_x, L_y] = i\hbar L_z$; there are evidently two further relations obtained by cyclic permutation of x, y, z. These commutation relations are satisfied by *any* angular momentum whatever—electron spin, for example (p. 71):

$$[\tfrac{1}{2}\hbar\sigma_x, \tfrac{1}{2}\hbar\sigma_y] = i\hbar(\tfrac{1}{2}\hbar\sigma_z).$$

The entire theory of quantum angular momentum can be built on these commutation relations.)

The nature of the energy spectrum

It is not possible to calculate the energy eigenvalues until the form of the potential $V(r)$ is given. However, there are certain important general features for *any* $V(r)$, which we shall consider now.

Suppose that, for each fixed l, we have been able to solve the one-dimensional χ-equation with the potential (p. 83)

$$v_l(r) = V(r) + \frac{\hbar^2 l(l+1)}{2m_0 r^2},$$

obtaining a set of eigenvalues and eigenfunctions E_{nl}, $\chi_{nl}(r)$, for every $l = 0, 1, 2, \ldots$ Then the *original* problem is solved by the set of functions

$$\psi_{nlm}(\boldsymbol{r}, t) = r^{-1}\chi_{nl}(r) Y_{lm}(\theta, \phi) \exp(-iE_{nl}t/\hbar),$$

each of which is simultaneously an eigenfunction of

H (energy) with eigenvalue E_{nl}

$\boldsymbol{L}^2$ (total squared angular momentum) with eigenvalue $\hbar^2 l(l+1)$

L_z (z-component of angular momentum) with eigenvalue $\hbar m$.

It is by now clear not only that separation of variables provides a convenient way of dealing with the problem of a spherically symmetric potential, but that the constants of separation represent important physical features of the stationary states.

The observables H, $\boldsymbol{L}^2$ and L_z are said to form a **complete commuting set**. They have to commute, since they share the same eigenfunctions (p. 26); they are said to be *complete* because two eigenfunctions with the same eigenvalues for H, $\boldsymbol{L}^2$, L_z differ at most by a phase factor. Consequently, the eigenvalues of H, $\boldsymbol{L}^2$, and L_z, *taken together*, may be used to *label* the stationary states of the system without ambiguity, and in a natural way.

The observable L_z cannot be left out of the set, even though the eigenvalue E_{nl} does not depend on m. In fact, since m may take any of $2l+1$ different values, there are $2l+1$ linearly independent simultaneous eigenvalues of H and $\boldsymbol{L}^2$. Any linear combination of these eigenfunctions is also an eigenfunction of both H and $\boldsymbol{L}^2$, but *not* of L_z: dropping L_z from the set allows a $(2l+1)$-fold degeneracy to occur. For this reason, the set of simultaneous eigenstates of H and $\boldsymbol{L}^2$ is often called a $(2l+1)$-plet, thus indicating the extent of the degeneracy. Spectroscopists have names of their own for these states, which have arisen in an *ad hoc* manner:

l	degeneracy	spectroscopic notation
0	singlet	s (sharp)
1	triplet	p (principal)
2	quintuplet (quintet)	d (diffuse)
3		f (fundamental)
	...	

(The names in the last column originally referred to observed series of spectral lines. The names triplet, etc., are used also to refer to clusters of spectral lines; the two uses must not be confused.)

As usual, we are particularly interested in the energy spectrum; that is, all the possible values of E_{nl}, whatever they may be. The nature of this energy spectrum is governed by the behaviour of $V(r)$ in much the same way as for the one-dimensional examples in chapter 5. For example, if $V(r) \to \infty$ as $r \to \infty$, then the same is true for the effective potential $v_l(r)$ for each l. Thus, for each l, the eigenvalue spectrum of the radial equation will consist of a *discrete* set of values, say

$$E_{1l}, E_{2l}, \ldots, E_{nl}, \ldots$$

which we may index with the integer n. The corresponding radial function is $\chi_{nl}(r)$, and may be normalized to satisfy

$$\int_0^\infty |\chi_{nl}(r)|^2 \, dr = 1.$$

Another case of considerable importance arises when $V(r) \to 0$ as $r \to \infty$. If $V(r) < 0$ for at least some values of r, it may be that there exists a discrete part to the spectrum with negative eigenvalues. In any event, any $E \geqslant 0$ is a possible eigenvalue. Such a case is the Coulomb electrostatic attraction

$$V(r) = -\frac{\text{const.}}{r};$$

in chapter 9 we shall see that this yields a spectrum whose discrete part comprises an *infinite* number of energy values: refer to the figure on p. 104.

The free particle

An obvious, if somewhat trivial, spherically symmetric potential is $V(r) = 0$ everywhere; this is the case of the *free particle*. Let us consider the solution in spherical polar coordinates—that is, let us find the simultaneous eigenstates of H, L^2, and L_z. These eigenstates will be different from the plane wave solutions found earlier (p. 64): these were eigenstates of the three components of linear momentum p_x, p_y, p_z.

The effective potential now involves only the centrifugal term,

$$v_l(r) = \frac{\hbar^2 l(l+1)}{2m_0 r^2}.$$

The radial equation (p. 83) is therefore

$$-\frac{\hbar^2}{2m_0}\frac{\mathrm{d}^2}{\mathrm{d}r^2}(rR) + \frac{\hbar^2 l(l+1)}{2m_0 r^2}.rR = E.rR,$$

the boundary condition at $r = 0$ being $rR = 0$. The solutions of this equation are

$$R_{kl} = j_l(kr) \quad \text{with} \quad \frac{\hbar^2 k^2}{2m_0} = E,$$

where the first few functions $j_l(\rho)$ are

$$j_0(\rho) = \frac{1}{\rho}\sin\rho,$$

$$j_1(\rho) = \frac{1}{\rho^2}\sin\rho - \frac{1}{\rho}\cos\rho,$$

$$j_2(\rho) = \left(\frac{3}{\rho^3} - \frac{1}{\rho}\right)\sin\rho - \frac{3}{\rho^2}\cos\rho.$$

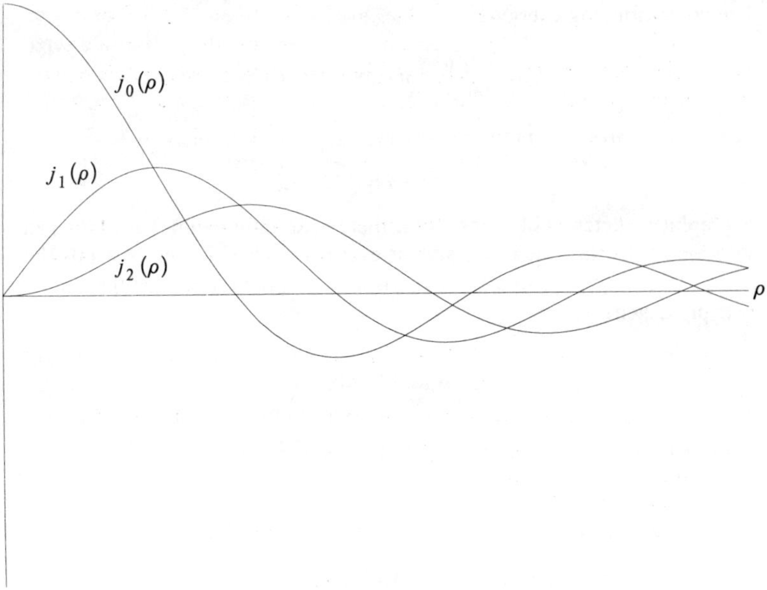

In general,

$$j_l(\rho) = (\text{polynomial in } \rho^{-1} \text{ of degree } l) \sin\rho$$
$$+ (\text{polynomial in } \rho^{-1} \text{ of degree } l-1) \cos\rho,$$

the polynomials being such that $j_l(\rho)$ tends to a finite limit as $\rho \to 0$. (Actually, $j_l(\rho)$ is the *spherical Bessel function* of order l.) The energy spectrum is continuous, since any value of k gives solutions which are bounded at $r = \infty$.

A similar system, where now the energy spectrum is discrete, is that of a particle contained in a spherical box, radius a, with a perfectly rigid wall. The radial equation does not change, but one boundary condition does; we now have

$$\psi(\mathbf{r}, t) = 0 \quad \text{at} \quad r = a.$$

In terms of the radial function, this requires that k must be chosen to satisfy—for a state with total angular momentum $\hbar l$—

$$j_l(ka) = 0.$$

For singlet states ($l = 0$) the solution is easy. The requirement $j_0(ka) = 0$ is simply $\sin ka = 0$, giving

$$ka = n\pi \quad \text{for} \quad n = 1, 2, 3, \ldots$$

The corresponding energy eigenvalues are

$$E_{n0} = \frac{\hbar^2\pi^2 n^2}{2m_0 a^2} \text{ for any integer } n > 0.$$

For triplet states ($l = 1$) the requirement is $j_1(ka) = 0$, or

$$\tan ka = ka.$$

A graphical sketch easily shows that there is an infinite number of discrete solutions k to this equation, almost every one of which gives a possible energy level E_{k1}. (The solution $k = 0$ is not acceptable.) Each of these levels is triply degenerate.

PROBLEMS

1. Reconsider problems 1 and 2 (p. 64) in the light of this section.
2. Solve $\tan ka = ka$ graphically. Why is $k = 0$ not acceptable?

9. The bound states of the hydrogen atom

Systems containing two particles

THE purpose of this chapter is to consider the behaviour of hydrogen-type atoms, for example, atomic hydrogen, deuterium, singly-ionized helium, doubly-ionized lithium, positronium, and so on. Such systems consist of two particles, which we shall call the *electron* (mass m_e, charge $-e$) and the *nucleus* (mass m_n, charge Ze). Any intrinsic structure of the particles, such as spin or magnetic moment, will be ignored for the moment.

The form of the wavefunction ψ_0 representing the state of such a system is easily found: imagine a simultaneous observation of the positions of the electron and the nucleus; for such an observation, the theory must provide a

$$\left.\begin{array}{l}\text{probability that the electron is found in a}\\ \text{volume } d^3\boldsymbol{r}_1 \text{ at position } \boldsymbol{r}_1 \text{ and the nucleus}\\ \text{is found in a volume } d^3\boldsymbol{r}_2 \text{ at position } \boldsymbol{r}_2\end{array}\right\} = |\psi_0(\boldsymbol{r}_1, \boldsymbol{r}_2, t)|^2 d^3\boldsymbol{r}_1 d^3\boldsymbol{r}_2,$$

by considerations which should be familiar by now. Thus $\psi_0(\ldots)$ is a function of $\boldsymbol{r}_1$, $\boldsymbol{r}_2$ and, of course, t. Once this is established, we may draw up a list of relevant observables, in the style of p. 62:

	electron	nucleus
mass	m_e	m_n
position	$\boldsymbol{r}_1 = (x_1, y_1, z_1)$	$\boldsymbol{r}_2 = (x_2, y_2, z_2)$
momentum	$-i\hbar\nabla_1 = -i\hbar\left(\dfrac{\partial}{\partial x_1}, \dfrac{\partial}{\partial y_1}, \dfrac{\partial}{\partial z_1}\right)$	$-i\hbar\nabla_2 = -i\hbar\left(\dfrac{\partial}{\partial x_2}, \dfrac{\partial}{\partial y_2}, \dfrac{\partial}{\partial z_2}\right)$
kinetic energy	$-\dfrac{\hbar^2}{2m_e}\nabla_1^2$	$-\dfrac{\hbar^2}{2m_n}\nabla_2^2$
mutual potential energy	$V(\lvert\boldsymbol{r}_1 - \boldsymbol{r}_2\rvert)$.	

We assume that the mutual potential energy depends only on the distance of separation $r = |\boldsymbol{r}_1 - \boldsymbol{r}_2|$; usually we shall have the Coulomb electrostatic potential in mind.

The Hamiltonian is the observable for the total energy, and this is here the sum of two kinetic energies and the mutual potential energy. It follows that the Schrödinger equation is

$$i\hbar\frac{\partial\psi_0}{\partial t} = -\frac{\hbar^2}{2m_e}\nabla_1^2\psi_0 - \frac{\hbar^2}{2m_n}\nabla_2^2\psi_0 + V(|\boldsymbol{r}_1 - \boldsymbol{r}_2|)\psi_0,$$

a partial differential equation for ψ_0 with seven independent variables.

Before we move on, two side-remarks may be pertinent. The first concerns feasibility. Any direct attack on a differential equation with more than a very few independent variables poses formidable problems. As a result, there has sprung up a branch of theoretical 'quantum chemistry' aimed at designing improved approximation techniques for the electronic states of atoms and molecules, often achieving notable success. Fortunately, we shall see below that the two-particle system has simplifying features.

The other remark concerns the nature of the wavefunction. The reader has probably already fallen into the temptation to think of ψ as representing some kind of 'smearing-out' of the quantum particle in space; indeed, some textbooks suggest that this is the right view to take. But it will not do. The function ψ_0 for the two-particle system is a function of *two* positions simultaneously, and such an entity is much more complicated than what would be needed merely to describe two smeared-out particles, namely, a pair of functions, each of one position only. In quantum mechanics, a composite system is incomparably greater than the mere sum of its parts. We shall discuss such matters a little more fully in chapter 18.

Separation into centre-of-mass and relative motions

In classical mechanics, a two-particle system may be handled by replacing the coordinates of the actual particles by the coordinates of the centre-of-mass and of relative position; in this way, a complete separation into two distinct problems may be achieved. The correspondence principle suggests that this may be possible for the analogous quantum system; we shall find that we are not disappointed.

Let us introduce new variables:

centre of mass position $\boldsymbol{R} = (m_e\boldsymbol{r}_1 + m_n\boldsymbol{r}_2)/(m_e + m_n)$, and

relative position $\boldsymbol{r} = \boldsymbol{r}_1 - \boldsymbol{r}_2.$

The x-components of these vectors are

$$X = (m_e x_1 + m_n x_2)/(m_e + m_n),$$

$$x = x_1 - x_2.$$

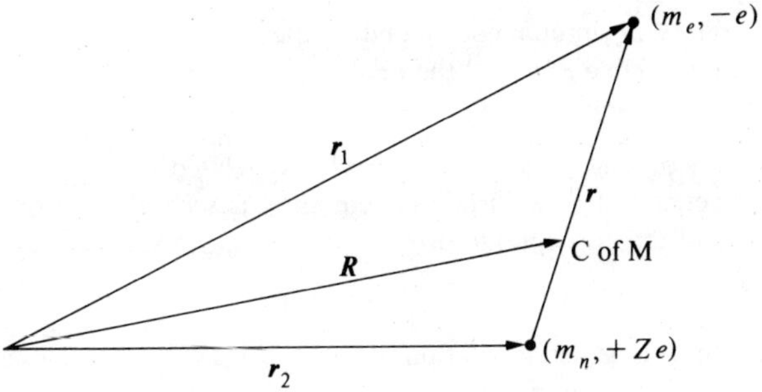

These relations lead to further relations between the partial derivatives with respect to the new variables, and those with respect to the old:

$$\frac{\partial}{\partial x_1} = \frac{\partial X}{\partial x_1} \cdot \frac{\partial}{\partial X} + \frac{\partial x}{\partial x_1} \cdot \frac{\partial}{\partial x} = \frac{m_e}{m_e + m_n} \frac{\partial}{\partial X} + \frac{\partial}{\partial x}$$

and similarly

$$\frac{\partial}{\partial x_2} = \frac{m_n}{m_e + m_n} \frac{\partial}{\partial X} - \frac{\partial}{\partial x}.$$

Hence

$$\frac{1}{m_e} \frac{\partial^2}{\partial x_1^2} + \frac{1}{m_n} \frac{\partial^2}{\partial x_2^2} = \frac{1}{m_e} \left(\frac{m_e}{m_e + m_n} \frac{\partial}{\partial X} + \frac{\partial}{\partial x} \right)^2 + \frac{1}{m_n} \left(\frac{m_n}{m_e + m_n} \frac{\partial}{\partial X} - \frac{\partial}{\partial x} \right)^2$$

$$= \frac{1}{M} \frac{\partial^2}{\partial x^2} + \frac{1}{\mu} \frac{\partial^2}{\partial x^2}$$

where

$$M = \text{total mass} \quad = m_e + m_n$$

$$\mu = \textbf{reduced mass} = \frac{m_e m_n}{m_e + m_n}.$$

Exactly similar considerations apply to the y- and z-coordinates, and we are led to

$$\frac{1}{m_e} \nabla_1^2 + \frac{1}{m_n} \nabla_2^2 = \frac{1}{M} \nabla_{\boldsymbol{R}}^2 + \frac{1}{\mu} \nabla_{\boldsymbol{r}}^2.$$

The Schrödinger equation may now be rewritten

$$i\hbar \frac{\partial \psi_1}{\partial t} = -\frac{\hbar^2}{2M} \nabla_{\boldsymbol{R}}^2 \psi_1 + \left\{ -\frac{\hbar^2}{2\mu} \nabla_{\boldsymbol{r}}^2 \psi_1 + V(r) \psi_1 \right\}.$$

The function $\psi_1(\boldsymbol{r}, \boldsymbol{R}, t)$ carries the meaning

$$\left.\begin{array}{l}\text{the probability of simultaneously finding the}\\ \text{c. of m. in a volume } d^3\boldsymbol{R} \text{ and the relative}\\ \text{position in a volume } d^3\boldsymbol{r}\end{array}\right\} = |\psi_1(\boldsymbol{r}, \boldsymbol{R}, t)|^2 d^3\boldsymbol{r} d^3\boldsymbol{R}.$$

(The functions ψ_0 and ψ_1 are obviously closely related.)

This last version of the Schrödinger equation is *separable*, so we search for solutions of the *very special* form

$$\psi_1(\boldsymbol{r}, \boldsymbol{R}, t) = \psi(\boldsymbol{r}, t)\Psi(\boldsymbol{R}, t),$$

knowing that the general solution (should we ever want it) may be expanded as a linear combination of such solutions. On substituting ψ_1 in the equation we find that each of Ψ and ψ must satisfy a Schrödinger equation of its own: the centre-of-mass equation

$$i\hbar \frac{\partial \Psi}{\partial t} = -\frac{\hbar^2}{2M} \nabla^2 \Psi, \qquad \text{(centre-of-mass motion)}$$

and the equation for relative motion

$$i\hbar \frac{\partial \psi}{\partial t} = -\frac{\hbar^2}{2\mu} \nabla^2 \psi + V(r)\psi, \qquad \text{(relative motion)}.$$

(We may now drop the suffixes on the symbol ∇ without ambiguity.) In this way the original problem has been replaced by the simultaneous solution of two completely independent and much simpler problems. Moreover, we have already considered them both. The centre-of-mass equation has exactly the form of the equation of a free particle of mass M; its treatment by plane waves is at p. 63, and by spherical harmonics at p. 88. The equation governing the relative motion is exactly the equation for a particle of reduced mass μ in a spherically symmetric potential $V(r)$, and this was the topic of Chapter 8.

PROBLEM

Carry out the separation of variables in detail.

The energy levels of a hydrogen atom

Clearly it is the equation for the *relative* motion which will lead to the characteristic properties of a hydrogen atom. The mutual potential energy in this case is the Coulomb potential

$$V(r) = \frac{-1}{4\pi\varepsilon_0} \frac{Ze^2}{r}.$$

(Including the atomic number Z allows us to treat other similar systems simultaneously; for hydrogen and deuterium $Z = 1$.) As on p. 86, we consider separated solutions

$$\psi = r^{-1}\chi_{nl}(r)Y_{lm}(\theta, \phi)\exp(-iE_{nl}t/\hbar)$$

where χ must satisfy

$$-\frac{\hbar^2}{2\mu}\frac{d^2\chi_{nl}}{dr^2} - \frac{1}{4\pi\varepsilon_0}\frac{Ze^2}{r}\chi_{nl} + \frac{\hbar^2 l(l+1)}{2\mu r^2}\chi_{nl} = E_{nl}\chi_{nl}$$

with

$$\chi_{nl}(0) = 0 \quad \text{and} \quad \chi_{nl}(r) \text{ bounded as } r \to \infty.$$

As discussed (p. 88), *any* positive E is a possible eigenvalue for this equation; the corresponding states are *scattering states* of the electron and the nucleus, rather than bound states of the atom. Henceforth we shall ignore them, and search for any discrete energy levels there may be. Such states will necessarily have $E < 0$, and it is convenient to write $E = -\hbar^2\lambda^2/2\mu$, and to search for the possible values of λ. The radial equation is now more conveniently written (dropping suffixes for the moment)

$$\frac{d^2\chi}{dr^2} + \frac{2Z}{\eta a_0 r}\chi - \frac{l(l+1)}{r^2}\chi = \lambda^2\chi,$$

introducing the length $a_0 = 4\pi\varepsilon_0\hbar^2/m_e e^2$, the **Bohr radius** of the hydrogen atom, and $\eta = m_e/\mu = 1 + m_e/m_n$, a correction factor to take into account the finite mass m_n of the nucleus.

Mathematical interlude

The radial equation evidently takes the form

$$\frac{d^2\chi}{dr^2} + \frac{A}{r}\chi - \frac{B}{r^2}\chi = \lambda^2\chi$$

where A, B, and λ are positive constants. Finding the bound states is equivalent to finding the relation between A, B, and λ which must hold for there to be a solution satisfying

$$\chi(0) = 0, \qquad \chi(r) \to 0 \quad \text{as} \quad r \to \infty.$$

It is convenient to deal with the problem with general coefficients A, B, as we shall later meet other equations of this form, and the results of this section will be used again then.

When r is large (that is, when the nucleus and electron are far apart), neglect of r^{-1} leads to the approximate equation

$$\frac{d^2\chi}{dr^2} = \lambda^2\chi,$$

whose solutions are obvious:

$$\chi = e^{\pm\lambda r}.$$

Thus for any acceptable bound state, we expect that χ will behave something like $e^{-\lambda r}$ as $r \to \infty$; the other sign is ruled out by the requirement that χ be bounded. This leads us to write

$$\chi = f(r)e^{-\lambda r}$$

and to switch our attention to the function f. We have

$$\frac{d^2\chi}{dr^2} = \left(\frac{d^2f}{dr^2} - 2\lambda\frac{df}{dr} + \lambda^2 f\right)e^{-\lambda r},$$

and substituting in the radial equation leads to an equation for f

$$\frac{d^2f}{dr^2} - 2\lambda\frac{df}{dr} + \frac{A}{r}f - \frac{B}{r^2}f = 0.$$

The boundary conditions are now

$$f(0) = 0, \qquad f(r) \text{ must } \textit{not} \text{ behave like } e^{2\lambda r} \text{ as } r \to \infty.$$

The simplest procedure is to expand f as a power series in r,

$$f = c_{s_o}r^{s_o} + c_{s_o+1}r^{s_o+1} + \ldots = \sum_{s_o} c_s r^s$$

where s_o, and the coefficients c, are to be found. (Note that s_o need not be an integer: the presence of r^{-1} and r^{-2} in the differential equation may result in f being singular at $r = 0$.) We substitute this series into the equation and collect the terms in a particular power of r (r^{s-1} in fact), and obtain

(1) $$s(s+1)c_{s+1} - 2\lambda s c_s + Ac_s - Bc_{s+1} = 0 \qquad \text{for } s \geqslant s_o$$

and (2) $$s_o(s_o-1)c_{s_o} - Bc_{s_o} = 0.$$

Since $c_{s_o} \neq 0$, equation (2) gives $s_o = \frac{1}{2} + (\frac{1}{4} + B)^{\frac{1}{2}}$. (The requirement $f(0) = 0$ rules out the other root.) The series for f therefore begins with a term in r^{s_o}, and the subsequent terms may be obtained by a repeated use of equation (1), in the form

$$c_{s+1} = \frac{2\lambda s - A}{s(s+1) - B}c_s.$$

It remains to ensure that f does not behave like $e^{2\lambda r}$ as $r \to \infty$. When s is large, and if we assume that no coefficient c_s is accidentally zero, then

$$\frac{c_{s+1}}{c_s} = \frac{2\lambda}{s} + 0\left(\frac{1}{s^2}\right),$$

which is similar to the ratio of successive coefficients in the power series expansion of $e^{2\lambda r}$. Then inevitably, f behaves like $e^{2\lambda r}$. Such disastrous behaviour can be avoided only if one of the coefficients c_s—and consequently every succeeding coefficient also—is zero. This is achieved only if the numerator $2\lambda s - A$ is zero for one of the values of s, that is, if $\lambda = A/2(s_o + n')$ for some integer $n' \geqslant 0$. When this happens, f behaves much more mildly as $r \to \infty$, being merely a polynomial: in fact, we now have a bound state.

To summarize: the values of λ which give bound states are

$$\lambda = A/2(s_o + n'), \quad \text{with} \quad s_o = \tfrac{1}{2} + (\tfrac{1}{4} + B)^{\frac{1}{2}}$$
$$\text{and} \quad n' \geqslant 0 \text{ is any integer.}$$

The corresponding radial function is

$$\chi(r) = e^{-\lambda r}.(c_{s_o}r^{s_o}+\ldots+c_{s_o+n'}r^{s_o+n'}),$$

where the coefficients c are related to their neighbours by

$$\{s(s+1)-B\}c_{s+1} = A(s_o+n')^{-1}(s-s_o-n')c_s.$$

(Note that if $n' = 0$, the polynomial contains just one term.) The relation between λ, A and B is found by eliminating s_o:

$$2\lambda\{n'+\tfrac{1}{2}+(\tfrac{1}{4}+B)^{\frac{1}{2}}\} = A.$$

This completes the search for bound states.

Digression: a particle under gravity

Later, we shall use a similar approach for the simple harmonic oscillator. It is possible that the reader may as a result come to believe that any sufficiently simple quantum system may have its bound states dealt with by the method of a terminating power series. This is not so. For example, the problem posed by

$$-\frac{\hbar^2}{2m}\frac{\partial^2\psi}{dx^2}+mgx\psi = E\psi$$

with

$$\psi(0) = 0, \quad \psi(x)\to 0 \quad \text{as} \quad x\to+\infty,$$

(i.e. the problem of a mass m confined to $x \geqslant 0$, acted on by a *constant* force $-mg$) *cannot* be solved in this way. The solution, though simple, involves a 'non-elementary' function (see problems). It is our good fortune that the results which are, from our point of view, *useful* may all be expressed in terms of 'elementary' functions (functions, that is, that involve nothing 'worse' than the exponential and the logarithm in their expression).

If the x-axis is taken to be vertical, the above equation describes the behaviour of a quantum particle under *gravity*. Sometimes it is felt that it is somehow 'wrong' that such a light object as an elementary particle should be affected by gravity; however, like any object with mass, a neutron in free fall, for example, will experience the usual acceleration g. The parabolic trajectory of slow neutrons from an atomic pile was first observed in 1951. There would be no objection in principle to putting a neutron in orbit round the earth, just like any satellite, though in practice it would be very vulnerable to environmental influences and in any case decay will almost certainly occur before even a single orbit is completed. In recent years, it has become possible to isolate extremely low-velocity ('ultracold') neutrons; their wavevectors are such that they are perfectly reflected from solid objects, and may therefore be conducted along pipes and stored in bottles! Such neutrons in a long enough vertical pipe—a few metres typically—will not make it to the top on account of gravity.

Of course, there is nothing particularly quantum-mechanical in all this; Newton's equations perform well enough. However, there are quantum-mechanical effects also to be observed. Consider a neutron diffraction experiment first carried out in 1975. An incoming horizontal monoenergetic neutron beam is diffracted at A, B, C, and D, by three lugs carved from a single silicon crystal; the angles of incidence and of diffraction are determined by the Bragg relation (p. 66). It is found that interference occurs at D. If the plane of the whole setup is vertical, then the neutron wavelength along CD is greater than that along AB (the neutrons lose momentum by climbing

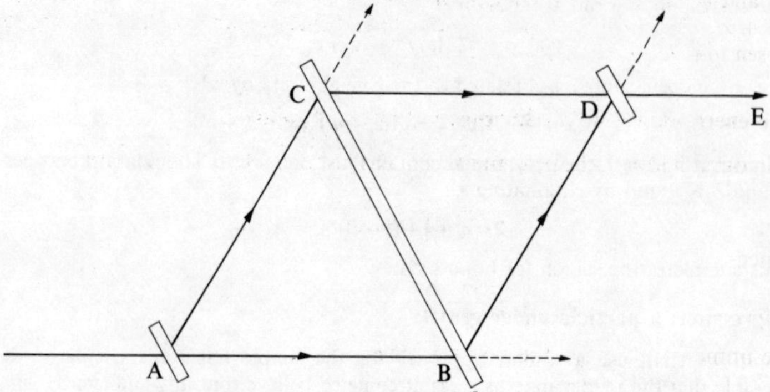

uphill along AC and along BD). If the plane is horizontal, the wavelengths are the same. Thus rotating the 'interferometer' about AB alters the phase relationship at D, leading to an intensity change in the emergent beam DE. The experiment is not easy, since the silicon crystal warps under its own weight, and needs to be kept rigid to about a lattice spacing: it has been likened to 'building an ordinary optical interferometer out of a blob of Jello'.

This effect is the **gravitational red-shift**. It was first observed in a laboratory experiment by Pound and Rebka in 1960, using the emission and absorption of gamma-photons from excited atomic nuclei.

PROBLEMS

1. Show that the substitutions $G^3 = 2m^2g/\hbar^2$, $\varepsilon = 2mE/\hbar^2$ lead to

$$-\frac{d^2\psi}{dx^2}+G^3x\psi = \varepsilon\psi, \qquad \psi(0)=0, \qquad \psi(x)\to 0 \quad \text{as} \quad x\to\infty.$$

2. Suppose the function $\phi(z)$ satisfies

$$-\frac{d^2\phi}{dz^2}+z\phi(z)=0, \qquad \phi(z)\to 0 \quad \text{as} \quad z\to\infty,$$

and that $\eta_1, \eta_2, \ldots$ are the zeros of ϕ, that is, $\phi(\eta_k)=0$. Show that the eigenvalues and eigenfunctions of the equation in problem 1 are

$$\varepsilon_k = -G^2\eta_k, \qquad \psi_k(x)=\phi(\eta_k+Gx).$$

3. Show that

$$\phi_1 = 1+\frac{z^3}{2.3}+\frac{z^6}{2.3.5.6}+\frac{z^9}{2.3.5.6.8.9}+\ldots$$

and

$$\phi_2 = z+\frac{z^4}{3.4}+\frac{z^7}{3.4.6.7}+\frac{z^{10}}{3.4.6.7.9.10}+\ldots$$

are both solutions of

$$-\frac{d^2\phi}{dz^2}+z\phi=0.$$

(ϕ_1 and ϕ_2 are related to Bessel functions of order $\pm\frac{1}{3}$. Both are unbounded as $z \to +\infty$. The function ϕ in problem 2 is a linear combination of ϕ_1 and ϕ_2, chosen to tend to zero as $z \to +\infty$.)

The energy levels of the hydrogen atom: conclusion

For our present problem, the relevant values of A, B (p. 95) are

$$A = 2Z/\eta a_0, \quad \text{and} \quad B = l(l+1);$$

hence

$$s_o = l+1 \quad \text{and} \quad \lambda = Z/\eta a_0(n'+l+1).$$

We immediately have the energy levels

$$E_{nl} = -\hbar^2\lambda^2/2\mu = -\frac{\hbar^2 Z^2}{2\eta m_e a_0^2}\cdot\frac{1}{n^2}$$

where $n = n'+l+1$ is any integer greater than l, and a_0 is the Bohr radius. The fact that, contrary to expectation, E_{nl} is independent of l will be discussed later.

The first few radial functions $R_{nl}(=\chi_{nl}/r)$ are easily determined up to a constant disposable factor with the help of the relation between neighbouring coefficients. The disposable factor may be chosen to ensure that

$$\int_0^\infty r^2\, \mathrm{d}r|R|^2 \equiv \int_0^\infty \mathrm{d}r|\chi|^2 = 1$$

in each case. When this is done, we find

$$R_{10}(r) = (Z/\eta a_0)^{3/2}\, 2\exp(-Zr/\eta a_0) \qquad (\text{for } E_{10} = -\hbar^2Z^2/2\eta m_e a_0^2)$$

$$R_{20}(r) = (Z/2\eta a_0)^{3/2}(2 - Zr/\eta a_0)\exp(-Zr/2\eta a_0) \quad (E_{20} = -\hbar^2Z^2/8\eta m_e a_0^2)$$

$$R_{21}(r) = (Z/2\eta a_0)^{3/2}(Zr/\eta a_0\sqrt{3})\exp(-Zr/2\eta a_0) \quad (E_{21} = E_{20})$$

etc.

These expressions for R_{nl} include polynomials of the dimensionless combination $Zr/\eta a_0$. They are related to the Laguerre (or Sonine) polynomials, and had arisen in a different connection in the nineteenth century, long before Schrödinger's solution for the hydrogen atom was developed.

Once the radial functions are known, the solution of the problem of the hydrogen atom bound states is completed by writing down the time-dependent separated normalized wavefunction (p. 95)

$$\psi_{nlm}(\mathbf{r}, t) = R_{nl}(r)Y_{lm}(\theta, \phi)\exp(-\mathrm{i}E_{nl}t/\hbar),$$

where, of course, the factors R and Y must share the same value for l.

PROBLEM

The results for the hydrogen atom may be 'scaled' in a very natural way. Assume throughout that $\eta = 1$, and define $\boldsymbol{\rho}$ and K through

$$a_0\boldsymbol{\rho} = \boldsymbol{r} \quad \text{and} \quad (\hbar^2/2m_e a_0^2)K = H.$$

In effect $\boldsymbol{\rho}$ measures electron position in units of the Bohr radius, while K relates the energy levels to the Rydberg constant R_∞; it measures energy in units of $chR_\infty = \hbar^2/2m_e a_0^2 = 2{\cdot}18 \times 10^{-10}$ J (or 13·6 eV), that is, the ionization energy of the hydrogen atom.

Show that

$$K = -\nabla_\rho^2 - \frac{2Z}{\rho}$$

and that typical eigenvalues and eigenfunctions of K are

nlm	eigenvalue K_n	eigenfunction $\psi(\rho, \theta, \phi)$
100	$-Z^2$	$(Z^3/\pi)^{\frac{1}{2}} \exp(-Z\rho)$
200	$-\frac{1}{4}Z^2$	$(Z^3/8\pi)^{\frac{1}{2}}(1-Z\rho/2) \exp(-Z\rho/2)$
210	$-\frac{1}{4}Z^2$	$(Z^3/32\pi)^{\frac{1}{2}} Z\rho \exp(-Z\rho/2) \cos\theta$
	etc.	

normalized so that in each case

$$\int d^3\boldsymbol{\rho}|\psi|^2 = 1.$$

(You will need to assemble information from pp. 81 and 99.)

Energy levels and spectroscopy

In this section we use the word *spectrum* in another, much older, sense. Many physical systems can emit or absorb electromagnetic radiation; the **emission** (or **absorption**) **spectrum** is then just the collection of possible frequencies (or, equivalently, wavelengths or wavenumbers) of such radiation. Such spectra have a long history. Newton in 1666 was possibly the first to suggest that the dispersion of sunlight by a glass prism into its constituents was a spectral analysis in some sense; in *what* sense, of course, he was unable to say. The presence of dark (i.e. absorption) **spectral lines** in sunlight was first observed by Wollaston in 1802, and the wavelengths of many of the stronger ones were measured by Fraunhofer twenty years later. Emission spectra from terrestrial sources were measured soon after, and soon a profusion of spectral lines had been catalogued.

It took a long time to detect any regularity in the spectral patterns of even the simplest systems, and it was nearly a century later when the **Ritz combination principle** was formulated (Rydberg 1900, Ritz 1908). This principle stated that the spectrum of any atomic or molecular system may be derived from a set of **spectral terms**, such that the wavenumbers of the lines of the spectrum are the differences of these terms taken in pairs.

It is now understood that the spectral terms (apart from a constant factor hc) are precisely the quantum energy levels of the system; thus spectroscopic observations have provided some of the most conclusive evidence of the reliability of modern quantum theory. In terms of the energy levels, the combination principle may be written as a conservation of energy:

$$h\nu = E_a - E_b,$$

where ν is the frequency of a spectral line, and E_a, E_b form a pair of levels. (This version is often known as the **Bohr frequency condition**.) The mechanism whereby two levels may jointly participate in this way will be discussed in chapter 15; it is closely connected with the way that the mean value of an observable changes in time (p. 33).

Optical spectroscopists tend to use *wavenumber* (that is, reciprocal wavelength) in their work; this is simply ν/c, where c is the velocity of light. Thus for the hydrogen atom, the spectral line wavenumbers are given by

$$\frac{\nu}{c} = \frac{R_\infty}{\eta}\left(\frac{1}{n_1^2} - \frac{1}{n_2^2}\right)$$

where both the **Rydberg constant** $R_\infty = \hbar/4\pi m_e a_0^2 c = 1{\cdot}097 \times 10^7\ \mathrm{m}^{-1}$ and, $\eta = 1 + m_e/m_n$ may be calculated from physical constants determined from experiments of a completely different kind. The agreement with observed spectra is excellent.

The spectrum of the deuterium atom

Deuterium (D) differs from hydrogen (H) only in the nature of its nucleus; the salient difference is that the mass of the nucleus is very nearly two proton masses. Consequently the *reduced* mass for D is different, and this has a slight, but easily observable, effect on the energy levels.

We have found

$$E_{nl} = -\frac{\hbar^2}{2\eta m_e a_0^2}\frac{1}{n^2}$$

($Z = 1$ for both H and D). For H, $\eta \sim 1{\cdot}000\,54$, while for D, $\eta \sim 1{\cdot}000\,27$. Thus the levels of H and D differ by a factor of about $1{\cdot}000\,27$; so therefore do the spectral wavenumbers. In fact, the existence of deuterium, though earlier suspected, was conclusively confirmed only in 1931 by a careful spectroscopic examination of atomic hydrogen. The usual H lines were found to have faint satellite lines at just the right distance, thus revealing the presence of a small amount of deuterium as impurity.

Singly ionised helium (He^+) was first detected in 1896 in the same kind of way, this time in the spectrum of a star, though it was believed at first that the lines were due to atomic hydrogen. For helium, $Z = 2$ and

$\eta \sim 1{\cdot}000\,14$; the levels are thus

$$E_{nl} = -\frac{\hbar^2}{2\eta m_e a_0^2}\frac{4}{n^2}.$$

The new factor 4 ensures that alternate levels are close to levels of H. It follows that many lines of He^+ will be found close to the lines of H. The correct interpretation of these lines was first given by Bohr.

The older quantum theory of Bohr

In 1913, Neils Bohr proposed a model of the hydrogen atom which was in its way a masterpiece in compromise. He imagined an electron rotating about a fixed nucleus, held in a circular orbit by the electrostatic attraction. The condition to be satisfied is that the 'centrifugal' and electrostatic forces must balance:

$$\frac{m_e v^2}{r} = \frac{Ze^2}{4\pi\varepsilon_0 r^2},$$

where v is the speed of the electron in its orbit. On the other hand, it had begun to be understood that if energy is quantized, so must the angular momentum of any periodic system be also (we have already met a modern version of such an argument on p. 21). Thus Bohr wrote down a second equation, with no classical foundation whatever:

$$\text{angular momentum} = m_e v r = n\hbar$$

where n is an *integer*. This is the quantum condition. These equations may be solved to give

$$v = \frac{Ze^2}{4\pi\varepsilon_0 n\hbar} \equiv \frac{Z\hbar}{n m_e a_0} \quad \text{and} \quad r = \frac{n\hbar}{m_e v} \equiv \frac{n^2 a_0}{Z}$$

where a_0 is the Bohr radius defined on p. 95. Then Bohr moved back to classical mechanics, and wrote the total energy as the sum of kinetic and potential energies:

$$E = \tfrac{1}{2}m_e v^2 - \frac{Ze^2}{4\pi\varepsilon_0 r} = -\frac{Z^2\hbar^2}{2m_e a_0^2}\frac{1}{n^2},$$

on substituting for v and r. The possible values of E are *exactly* those we have already determined via the Schrödinger equation.

Bohr's work was a milestone for the understanding of atomic spectra. It predicted a value for the Rydberg constant which, when corrected for the effect of reduced mass, was very close to the experimental value. Moreover, the model gained further credence on account of its behaviour for large n. Suppose we consider a value of n about, say, a million. Then $n^2 a_0$ would be about 53 m, and the hydrogen atom would be enormous. (Of course, any laboratory investigation of such an atom is impracticable! We must be content to think about it.) By the correspondence principle (and this is where Bohr's correspondence principle first appeared), the frequency of radiation of such a system may be calculated in two ways, which must agree because the system is macroscopic. Classically, the radiation frequency must be simply the frequency of the electron's rotation in its orbit, namely

$$\nu = \frac{v}{2\pi r} = \frac{\hbar}{2\pi m_e a_0^2} \frac{1}{n^3} \qquad (Z = 1)$$

(which, for $n \sim 10^6$, is a very leisurely 0·007 Hz; the orbital period is about $2\frac{1}{2}$ minutes). Quantum mechanically, we calculate

$$\nu = \frac{1}{2\pi\hbar}(E_n - E_{n-1}) = -\frac{\hbar}{4\pi m_e a_0^2}\left\{\frac{1}{n^2} - \frac{1}{(n-1)^2}\right\}$$

$$= +\frac{\hbar}{2\pi m_e a_0^2}\left\{\frac{1}{n^3} + 0\left(\frac{1}{n^4}\right)\right\}.$$

The two results agree to a few parts in n. Thus for large n, the quantum and classical predictions agree.

All this is very persuasive. It is good to know about Bohr's model: not only did it point the way to go at the time, but it was also a valuable object-lesson in how to go to the heart of the matter, retaining what is relevant and rejecting what is not. However, we now know that in spite of its incontrovertibly correct result for energy, Bohr's theory is wrong in one fatal particular: the angular momentum of the nth level is predicted to be $n\hbar$, and this is just what it is not. As we have seen, the possible values are in fact $0, \hbar, 2\hbar, \ldots (n-1)\hbar$, but not $n\hbar$, and this shows up a shortcoming fundamental in a theory which relies so heavily for its success on the quantization of angular momentum. This needs to be said, because more than half a century later the Bohr model is still occasionally offered as an approach which is claimed to be equivalent to the Schrödinger formulation, and is to be preferred for students who are unwilling to tackle the more demanding mathematics! In this connection, a reminiscence of Felix Bloch is worth quoting:

> Once at the end of a colloquium I heard Debye saying something like: 'Schrödinger, you are not working right now on very important problems anyway. Why don't you tell us some time about that thesis of de Broglie, which seems to have attracted some attention.'
>
> So, in one of the next colloquia, Schrödinger gave a beautifully clear account of how de Broglie associated a wave with a particle and how he could obtain the quantization rules of Niels Bohr and Sommerfeld by demanding that an integer

number of waves should be fitted along a stationary orbit. When he had finished, Debye casually remarked that he thought this way of talking was rather childish. As a student of Sommerfeld he had learned that, to deal properly with waves, one had to have a wave equation. It sounded quite trivial and did not seem to make a great impression, but Schrödinger evidently thought a bit more about the idea afterwards.

Just a few weeks later he gave another talk in the colloquium which he started by saying: 'My colleague Debye suggested that one should have a wave equation; well, I have found one!'

(*Physics Today*/December 1976, p. 23.)

Degeneracy

The expression for E_{nl} (p. 99) has turned out to be *independent* of l. This phenomenon does not occur in general, and this makes the Coulomb potential very special. One of the most important consequences is that the energy levels have an unusually high degeneracy, and the number of spectroscopic terms is greatly reduced, making the spectrum of a hydrogen atom easier to analyse than it might have been.

Let us determine the degeneracy of any chosen level. The degeneracy of those states for which both n and l are fixed is $2l+1$, this being the number of possible m values with $|m| \leqslant l$ (p. 81). For any fixed n, l is allowed to take any value in $0 \leqslant l < n$, and we have now found that all the levels E_{n0}, $E_{n1}, \ldots, E_{n,n-1}$ are equal. Their degeneracies are respectively $1, 3, \ldots, 2n-1$, giving a *total* degeneracy of

$$1+3+\ldots+(2n-1) = n^2,$$

as illustrated.

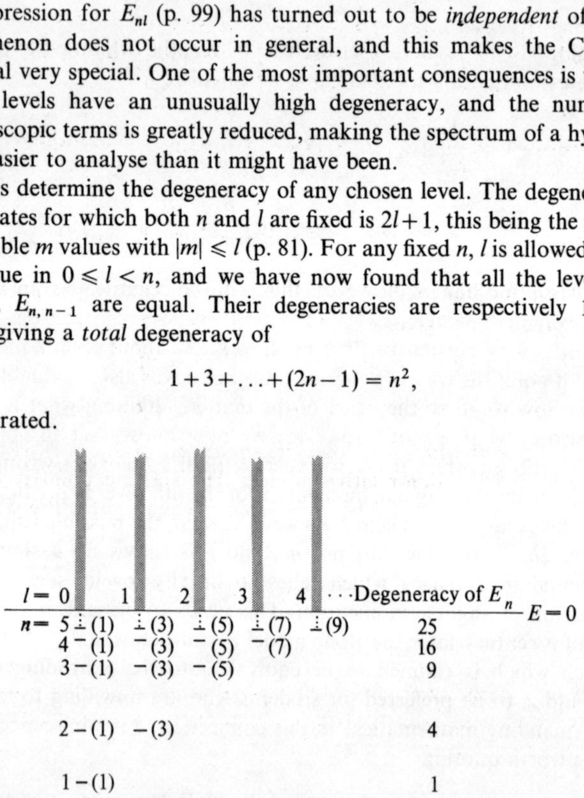

There are ways of reducing this degeneracy. All we have to do is to apply some kind of external influence which will alter the Hamiltonian enough to allow levels, previously equal, to become different. We then say that the degeneracy has been **lifted**, or that the level has been **split**. Any such splitting will result in a corresponding splitting in the observed spectral lines, thus providing a powerful tool for the investigation of atomic structures.

The most commonly applied external influences are constant electric and magnetic fields, which yield splittings which are known respectively as the **Stark** and **Zeeman effects**. These will be considered in Chapter 13.

Fine-structure

Careful experiment shows that even in the absence of any external influence, the energy levels of the hydrogen atom are not quite degenerate; the 'lines' in the spectrum are not single lines at all, but clusters of closely-spaced lines forming a **fine structure**. It is possible to account for this fine structure by taking *relativistic corrections* into consideration, and as this is straightforward to do exactly and forms an important test of the theory, we shall do it now.

In classical mechanics, conservation of energy for the motion of a mass μ in a spherically symmetric V may be written

$$E = \frac{1}{2\mu}\boldsymbol{p}^2 + V(r) \qquad (\boldsymbol{p} = \text{particle momentum}).$$

In *relativistic* mechanics, however, this is to be regarded as an approximation (when c^{-2} is neglected) to the more nearly exact relation

$$(\mathscr{E} - V)^2 = \mu^2 c^4 + c^2 \boldsymbol{p}^2 \qquad (c = \text{velocity of light}$$
$$\mathscr{E} = \mu c^2 + E).$$

This is all the relativity we need to know; the rest of the work is basically the same as for the non-relativistic case. The starting point is to write down, by analogy, a relativistic 'Schrödinger' equation for the stationary states of energy $\mathscr{E}$,

$$(\mathscr{E} - V)^2 \psi = \mu^2 c^4 \psi - c^2 \hbar^2 \nabla^2 \psi,$$

or

$$c^2 \hbar^2 \nabla^2 \psi + (V^2 - 2\mathscr{E} V)\psi = (\mu^2 c^4 - \mathscr{E}^2)\psi.$$

The only new feature is the algebraic inconvenience of having $\mathscr{E}$ appearing in more than one term. This equation is separable; we write $\psi = r^{-1}\chi(r)Y(\theta, \phi)$ as before, and find

$$\frac{d^2\chi}{dr^2} - \frac{l(l+1)}{r^2}\chi + \frac{1}{c^2\hbar^2}(V^2 - 2\mathscr{E} V)\chi = \left(\frac{\mu^2 c^2}{\hbar^2} - \frac{\mathscr{E}^2}{c^2\hbar^2}\right)\chi.$$

This equation is to be solved for the Coulomb potential

$$V = -\frac{Ze^2}{4\pi\varepsilon_0 r} = -\frac{Z\alpha\hbar c}{r},$$

where we have introduced the important dimensionless **fine structure constant**

$$\alpha = \frac{e^2}{4\pi\varepsilon_0\hbar c} = 0{\cdot}0073 \simeq \frac{1}{137}.$$

The equation for χ now becomes

$$\frac{d^2\chi}{dr^2} - \frac{l(l+1)}{r^2}\chi + \left(\frac{Z^2\alpha^2}{r^2} + \frac{Z\alpha\mathscr{E}}{c\hbar r}\right)\chi = \lambda^2\chi.$$

This equation is exactly of the form we have already dealt with (p. 95), where now

$$A = Z\alpha\mathscr{E}/c\hbar \quad \text{and} \quad B = l(l+1) - Z^2\alpha^2,$$

and

$$\lambda^2 = \frac{\mu^2c^2}{\hbar^2} - \frac{\mathscr{E}^2}{c^2\hbar^2}.$$

There is no need to repeat any of the work. The relation on p. 97 gives in this case

$$2\lambda\{n' + \tfrac{1}{2} + (\tfrac{1}{4} + l(l+1) - Z^2\alpha^2)^{\frac{1}{2}}\} = Z\alpha\mathscr{E}/c\hbar.$$

We use the last two relations to eliminate λ, obtaining the *energy levels*

$$\mathscr{E}_{n'l} = \mu c^2\left(1 + \frac{Z^2\alpha^2}{(n' + \frac{1}{2} + [(l + \frac{1}{2})^2 - Z^2\alpha^2]^{\frac{1}{2}})^2}\right)^{-\frac{1}{2}}.$$

This is an *exact* result. To second order in α,

$$\mathscr{E}_{n'l} = \mu c^2 + E_n\left\{1 + \frac{Z^2\alpha^2}{n^2}\left(\frac{n}{l+\frac{1}{2}} - \frac{3}{4}\right)\right\}$$

where $n = n' + l + 1 (> l)$, showing that the n^2-degeneracy is not lifted until terms of order α^2 are reached. Since α^2 is small (about 0·000053) it is to be expected that the splitting will be small. The presence of Z^2 shows that (for example) the effect for He^+ will be much more prominent than it is for H.

Electron spin

The exciting feature of the theoretical prediction for the fine structure for hydrogen is that agreement with experiment is poor! A qualitative reason for this is that (classically) a particle moving through an electric field will experience an induced magnetic field. If the particle possesses a magnetic moment (as the electron does), the resulting interaction may affect the observations.

This is not the place to consider the relativistic theory of a particle of spin $\frac{1}{2}\hbar$. Such a theory was developed by Dirac in 1928, and is by no means straightforward. However, two exact predictions are worth mentioning: the first is the electron magnetic moment should be $e\hbar/2m_e$ almost exactly ($\mu_B \equiv e\hbar/2m_e$ is the **Bohr magneton**, with the value $9{\cdot}274 \times 10^{-24}$ JT^{-1}). This is well substantiated experimentally. The other prediction concerns fine structure. The Dirac equation with a Coulomb potential is exactly soluble, and the energies of the bound states turn out to be

$$\mathscr{E}_{n'k} = \mu c^2\left(1+\frac{Z^2\alpha^2}{(n'+\sqrt{\{k^2-Z^2\alpha^2\}})^2}\right)^{-\frac{1}{2}}$$

where the quantum numbers n', k are integers, $n' \geqslant 0$ and $k \geqslant 1$. This result is in excellent agreement with observation. We have here a powerful confirmation of many aspects of quantum mechanics and of relativity, along with a convincing, though rather indirect, demonstration that the electron spin is indeed $\frac{1}{2}\hbar$.

To second order in α,

$$\mathscr{E}_{n'k} = \mu c^2 + E_n\left(1+\frac{Z^2\alpha^2}{n^2}\left(\frac{n}{k}-\frac{3}{4}\right)\right), \qquad (n = n'+k).$$

The only difference from the formula without spin is the absence of the $\frac{1}{2}$ in a denominator. Curiously, 13 years earlier, Sommerfeld had extended Bohr's work to include *elliptical* orbits with relativistic corrections, which he quantized according to the rules of the time. The formula he obtained was this one, which 'explained' the fine structure very well indeed. Later it was discovered that the two corrections that were necessary—the passage to a more adequate quantum theory, and the inclusion of electron spin—cancel each other more or less exactly as far as energy levels are concerned. As with Bohr's theory, the predicted angular momenta were wrong.

PROBLEMS

1. For fixed n, show that the values of

$$E_n\left(1+\frac{Z^2\alpha^2}{n^2}\left(\frac{n}{l+\frac{1}{2}}-\frac{3}{4}\right)\right)$$

vary over a range $E_nZ^2\alpha^2 . 4(n-1)/n(2n-1)$ as l goes from 0 to $n-1$.

2. For fixed n, show that the values of

$$E_n\left(1+\frac{Z^2\alpha^2}{n^2}\left(\frac{n}{k}-\frac{3}{4}\right)\right)$$

vary over $E_nZ^2\alpha^2 . (n-1)/n^2$ as k goes from 1 to n.

(These results serve to distinguish the two cases spectroscopically. The following table is given for the sake of comparison:

n	$4(n-1)/n(2n-1)$	$(n-1)/n^2$
1	0	0
2	0·67	0·25
3	0·53	0·22
4	0·43	0·19
5	0·36	0·16
6	0·30	0·14

The last column corresponds to the experimentally observed effect.)

Recoil and Doppler broadening

So far we have ignored any effect that the centre-of-mass motion may have on the emission spectrum of a hydrogen atom. However, a photon of energy $\hbar\omega$ carries a linear momentum $\hbar\omega/c$, and we ought to consider the conservation of momentum in addition to energy.

For our purposes, it is sufficient to use the classical conservation laws. We consider the following process: the atom initially has a velocity $\boldsymbol{v}_a$, and is in an excited level E_a; a photon of energy $h\nu$ and momentum $(0, 0, h\nu/c)$ is emitted; the atom is left with velocity $\boldsymbol{v}_b$ and is in a different level E_b. Conservation of energy (i.e. the Bohr frequency condition) is

$$h\nu = (E_a + \tfrac{1}{2}M\boldsymbol{v}_a^2) - (E_b + \tfrac{1}{2}M\boldsymbol{v}_b^2) = \Delta E + \tfrac{1}{2}M(\boldsymbol{v}_a^2 - \boldsymbol{v}_b^2)$$

where $\Delta E = E_a - E_b$. Conservation of momentum gives

$$(0, 0, h\nu/c) = M\boldsymbol{v}_a - M\boldsymbol{v}_b.$$

Together, these relations give

$$\begin{aligned} h\nu &= \Delta E + \tfrac{1}{2}M(\boldsymbol{v}_a + \boldsymbol{v}_b) \, . \, (0, 0, h\nu/Mc) \\ &= \Delta E - h\nu V/c \end{aligned}$$

where $-V$ is the z-component of $\frac{1}{2}(\boldsymbol{v}_a + \boldsymbol{v}_b)$. Hence

$$h\nu = \frac{\Delta E}{1 + V/c},$$

showing that the possibility of a centre-of-mass motion is fully accounted for by a simple correction factor.

Suppose now that the atom is initially at rest: $\boldsymbol{v}_a = 0$. Then

$$|\boldsymbol{v}_b| = h\nu/Mc, \quad \text{and} \quad V = h\nu/2Mc.$$

Hence

$$h\nu = \frac{\Delta E}{1 + h\nu/2Mc^2}.$$

Now $h\nu$ is the energy of a photon in the *visible* spectrum, or thereabouts, and is typically a few electron-volts. Mc^2, on the other hand, is the relativistic rest-energy of a hydrogen atom, about 10^9 electron-volts. Thus the correction factor has an utterly negligible effect, and may be ignored. The *recoil* of the atom is thus entirely unimportant.

If the atom is not initially at rest, the correction factor, though still small in most situations, produces an observable effect: the line experiences a Doppler-shift by an amount related to the velocity v of the atom in the direction of propagation of the photon. If, as is usual, the spectrum is emitted by an *aggregate* of particles possessing a distribution of velocities (a hot gas, say), the line appears to be slightly smeared out; this is the phenomenon of **Doppler broadening**.

Spectral lines may be broadened in other ways also. In fact, every line has a *natural linewidth*, which is roughly inversely proportional to the time taken by the system to emit a photon. The reasons for this linewidth will be elucidated by the more complete theory of radiation in Chapter 15.

10. The Dirac notation

Why introduce a new notation?

WHY indeed? We have managed quite well up to now with the notation we already have; why complicate matters? There are some who may suggest that Dirac developed his notation merely because he liked elegance. However, no technique gains general acceptance unless it is a genuinely useful tool, and the Dirac notation is exactly that. Though it would be perfectly possible to continue without learning a new notation, there are several reasons why this would be to our disadvantage.

The first reason is *conciseness*: a good compact notation can help a mathematical development along enormously. (Think of ordinary three-dimensional vector notation; for example, verify the simple vector identity

$$\boldsymbol{a} \wedge (\boldsymbol{b} \wedge \boldsymbol{c}) = (\boldsymbol{a} \cdot \boldsymbol{c})\boldsymbol{b} - (\boldsymbol{a} \cdot \boldsymbol{b})\boldsymbol{c}$$

by writing out each side in its full component form.) Second, there is *flexibility*: a good notation may give us more freedom to change direction when appropriate. (Again from vector notation, some problems are most easily set up in one system of coordinates—or even directly in vector form—and most easily solved in another system. It is therefore useful to be able to change coordinates at will.) Thirdly, *generality*: there are certain results in quantum mechanics which are of universal application, and it pays to use the general notation to arrive at these results in the form of recipes which may then be applied in particular cases.

On account of all this, we are not surprised to find a fourth reason for coming to terms with the Dirac notation: most people use it, and it is desirable to know what they are trying to say.

Setting up the Dirac notation for a particle in one dimension

Consider first an analogy, to illustrate the kind of thing we are about to do. In ordinary space of three dimensions, the components of a vector $\boldsymbol{r}$ in a given Cartesian frame are (x, y, z). We may write

$$x = \mathbf{i} \cdot \boldsymbol{r}, \qquad y = \mathbf{j} \cdot \boldsymbol{r}, \qquad z = \mathbf{k} \cdot \boldsymbol{r}$$

where the *basis vectors* $\mathbf{i}, \mathbf{j}, \mathbf{k}$ are unit vectors along the three coordinate axes. Furthermore, $\boldsymbol{r}$ may be reconstituted from its components by

$$\boldsymbol{r} = x\mathbf{i} + y\mathbf{j} + z\mathbf{k},$$

again with the help of the basis vectors. Similar relations between vector, components and basis appear—with appropriate modifications—in quantum mechanics, and form the foundation of the Dirac notation.

To fix ideas, we shall do the work for a particle in one dimension, whose state is represented by a wavefunction $\psi(x,t)$. (The details are different, sometimes *very* different, for other systems, but the basic ideas are exactly the same.) For the time being, we drop all reference to the time t, as all the work is carried out for one time only. Up to now, states Ψ have been represented by wavefunctions according to the scheme

$$\Psi \leftarrow \begin{cases} \psi(x) \\ \textit{or} \\ \psi^*(x) \end{cases}.$$

(The backward arrow $\leftarrow$ may be read 'is represented by'.) It is now time to introduce two new entities, the **statevectors** $|\Psi\rangle$ and $\langle\Psi|$, either of which may be used to represent the state Ψ, and each of which is represented in its turn by its 'components', which are in fact the various values taken by ψ (or ψ^*) as x goes from $-\infty$ to ∞. We therefore amplify the scheme to

$$\Psi \leftarrow \begin{cases} |\Psi\rangle \leftarrow \psi(x) \\ \textit{or} \\ \langle\Psi| \leftarrow \psi^*(x) \end{cases}.$$

Two statevectors are needed; this is because a state may be represented (i) by a column vector or by the Hermitian conjugate row vector (if the system is simple enough, p. 7), or (ii) by $\psi(x)$ or $\psi^*(x)$ (as in the case now under consideration). The added complication that each statevector has an *infinite* number of 'components' (one value $\psi(x)$ for each value of x) is special to the system we are considering, namely, the particle in one dimension. The number of components in any particular case will always depend on the structure of the system in question.

Two statevectors, one of each kind, may be combined to form a **scalar product**

$$\langle\Phi|\,.\,|\Psi\rangle \equiv \langle\Phi|\Psi\rangle,$$

whose value is always a number. Conventionally, the style on the right is always followed where possible, all unnecessary vertical bars being omitted. (Dirac chose his notation for statevectors so that the scalar product should look like a rather distinctive *bracket expression*. The statevectors themselves are thus *incomplete* bracket expressions, and were called by Dirac **bra** vectors and **ket** vectors, from the happy circumstance that the word 'bracket' has two syllables. These names are still occasionally used.)

Insofar as, for each x, $\psi(x)$ is a 'component' of the statevector $|\Psi\rangle$, it ought to be possible to write this $\psi(x)$ as a scalar product of a **basis vector** (which we shall denote by $\langle x|$, using an obvious labelling) with the statevector $|\Psi\rangle$,

$$\psi(x) = \langle x|\Psi\rangle.$$

(Compare with, for example, $z = \mathbf{k}\,.\,\boldsymbol{r}$.) The set of all the basis vectors $\langle x|$, one for each x, is called the **basis** of the x-representation (for reasons which will become apparent in due course); this basis is analogous to the set **i, j, k** in ordinary vector algebra. Moreover, we need to be able to reconstitute $|\Psi\rangle$ from a knowledge of its 'components' $\psi(x)$; this is done by writing the **superposition**

$$|\Psi\rangle = \int_{-\infty}^{\infty} |x\rangle\, \mathrm{d}x \psi(x),$$

by analogy with $\boldsymbol{r} = \mathbf{i}x + \mathbf{j}y + \mathbf{k}z$. (We expect an integral rather than the mere sum of three terms, since there is a continuous infinity of 'components', each of which makes its own contribution.) We have here introduced a *second* set of basis vectors $|x\rangle$. The nature of the physical state represented by $|x\rangle$ (or $\langle x|$) will be considered later.

We may combine the last two equations to give

$$\begin{aligned} |\Psi\rangle &= \int_{-\infty}^{\infty} |x\rangle\, \mathrm{d}x \langle x|\Psi\rangle \\ &= \left\{ \int_{-\infty}^{\infty} |x\rangle\, \mathrm{d}x \langle x| \right\} |\Psi\rangle. \end{aligned}$$

(We *always* assume that factors like $|\Psi\rangle$ may legitimately be moved in and out of integrals.) As this relation is true for any state Ψ, we obtain the very important result that the expression in $\{\ldots\}$ is the **unit operator**,

$$\int_{-\infty}^{\infty} |x\rangle\, \mathrm{d}x \langle x| = 1.$$

(This may be compared with the eigenvalue–eigenvector expansion of a unit matrix, p. 16.)

A frequently useful algebraic device is to insert a unit operator at appropriate points in an expression (this has no effect), and then to replace it with the formula just given. For example, consider the scalar product of $\langle\Phi|$ and $|\Psi\rangle$; inserting a unit operator leads to

$$\begin{aligned} \langle\Phi|\Psi\rangle &= \langle\Phi| \int_{-\infty}^{\infty} |x\rangle\, \mathrm{d}x \langle x|\, |\Psi\rangle \\ &= \int_{-\infty}^{\infty} \langle\Phi|x\rangle\, \mathrm{d}x \langle x|\Psi\rangle. \end{aligned}$$

(In the last step, both $\langle\Phi|$ and $|\Psi\rangle$ are moved into the integrand.) Now if Ψ and Φ are represented respectively by $\psi(x)$ and $\phi(x)$, then the correct scalar product is (p. 35)

$$(\phi^*, \psi) = \int_{-\infty}^{\infty} \phi^*(x)\psi(x)\,\mathrm{d}x.$$

Hence, since $\psi(x) = \langle x|\Psi\rangle$, the expressions agree if

$$\langle\Phi|x\rangle = \phi^*(x) = (\langle x|\Phi\rangle)^*.$$

This is a special case of a now easily proved general result for *any* two state vectors

$$\langle\Psi|\Phi\rangle = (\langle\Phi|\Psi\rangle)^*.$$

The *Hermitian conjugate* of

$$|\Psi\rangle = \int_{-\infty}^{\infty} |x\rangle\,\mathrm{d}x\psi(x)$$

is defined as

$$\langle\Psi| = \int_{-\infty}^{\infty} \psi^*(x)\,\mathrm{d}x\langle x|.$$

Consequently,

$$\langle\Psi|\Psi\rangle = \int_{-\infty}^{\infty} \psi^*(x)\,\mathrm{d}x\langle x|\Psi\rangle$$

$$= \int_{-\infty}^{\infty} |\psi(x)|^2\,\mathrm{d}x;$$

and the condition that the statevector $|\Psi\rangle$ is *normalized* is thus

$$\langle\Psi|\Psi\rangle = 1.$$

The Dirac δ-function

Consider the effect of inserting a unit operator in $\langle x|\Psi\rangle$,

$$\psi(x) = \langle x|\Psi\rangle = \langle x|\int_{-\infty}^{\infty} |x'\rangle\,\mathrm{d}x'\langle x'|\,|\Psi\rangle$$

$$= \int_{-\infty}^{\infty} \langle x|x'\rangle\,\mathrm{d}x'\psi(x').$$

The left side is the value of ψ at the point x, while the right side apparently involves values at other points $x' \neq x$. This cannot be so, and we deduce that

$$\langle x|x'\rangle = 0 \quad \text{unless} \quad x = x'.$$

This cannot be any ordinary function of x and x', as it is to be zero almost everywhere.

To cope with this situation, Dirac introduced an *improper* function $\delta(x)$ (the **Dirac delta-function**) with the properties

$$\int_{-\infty}^{\infty} \delta(x)\,\mathrm{d}x = 1,$$

$$\delta(x) = 0 \quad \text{for} \quad x \neq 0.$$

Obviously, $\delta(x)$ is not a function of x in any usual sense; nevertheless, there exists a complete and fully respectable *distribution theory* of such 'improper' functions, and it is possible to handle $\delta(x)$ as if it were an ordinary function for all our purposes without obtaining wrong results. In particular, it is guaranteed that under all normal circumstances any δ-function which appears will ultimately form part of an *integrand*, and thus disappear from any expression relating directly to the result of physical experiment. To know this should be sufficient to reassure us.

To obtain an intuitive picture of $\delta(x)$ imagine a function which vanishes everywhere except inside a small interval of length ε surrounding the origin $x = 0$, and which is so large in this interval that its integral is unity. For example,

$$f(x;\varepsilon) = \begin{cases} 1/\varepsilon & \text{for} \quad -\varepsilon/2 \leqslant x \leqslant \varepsilon/2 \\ 0 & \text{elsewhere} \end{cases}$$

will do (figure (a)). As $\varepsilon \to 0$, $f(x;\varepsilon)$ gives an ever-improving idea of what $\delta(x)$ must look like.

The most important property of $\delta(x)$ is

$$\int_{-\infty}^{\infty} f(x)\delta(x)\,\mathrm{d}x = f(0),$$

and, somewhat more generally,

$$\int_{-\infty}^{\infty} f(x)\delta(x-a)\,\mathrm{d}x = f(a).$$

From this, we can now see that it is correct to write

$$\langle x|x'\rangle = \delta(x-x').$$

A purist may wish to develop the whole of quantum mechanics without ever introducing the δ-function. Such a procedure, though perfectly possible, leads to a grave loss of simplicity, and is now never considered. It

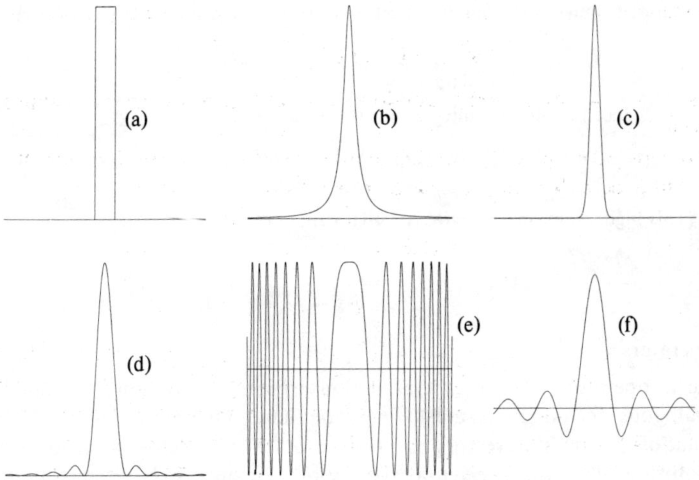

was not always so: the δ-function had already appeared in various guises in the nineteenth century, particularly in the work of Heaviside, and had often been attacked on the grounds that it was thoroughly unrespectable. The usual view now is that, like the Dirac notation itself, it is a useful tool that it would be a pity to discard.

PROBLEMS

1. In $\int f(x)\delta(x)\,\mathrm{d}x$, change the variable of integration to $y = y(x)$. Hence show that

$$\delta(y(x) - y(0)) = \delta(x) \Big/ \left[\frac{\mathrm{d}y}{\mathrm{d}x}\right]_{x=0}.$$

2. Show that

(i) $\displaystyle\int_{-\infty}^{\infty} \frac{a}{\pi(x^2+a^2)}\,\mathrm{d}x = 1;$

(ii) as $a \to 0+$, the maximum of the function $\dfrac{a}{\pi(x^2+a^2)}$ becomes higher and narrower.

Deduce a reasonable 'approximation' to $\delta(x)$. (See figure (b).)

3. Consider similarly the functions

(i) $\pi^{-\frac{1}{2}}\lambda \exp(-\lambda^2 x^2)$ as $\lambda \to +\infty$ (figure (c))

(ii) $(1 - \cos \alpha x)/\pi\alpha x^2$ as $\alpha \to +\infty$ (figure (d))

(iii) $(2/\pi)^{\frac{1}{2}}\lambda \cos \lambda^2 x^2$ as $\lambda \to +\infty$ (figure (e)).

(It quite often happens that a function with some such shape arises in the course of the development, and that replacement with a δ-function may be an excellent approximation. See p. 184 for an important example.)

4. Consider similarly the function $(\pi x)^{-1} \sin ax$ as $a \to \infty$ (figure (f)). Show that

$$\frac{1}{2\pi}\int_{-a}^{a} e^{ikx}\,dk = (\pi x)^{-1} \sin ax,$$

and deduce the 'Fourier integral' for the Dirac function

$$\delta(x) = \frac{1}{2\pi}\int_{-\infty}^{\infty} e^{ikx}\,dk.$$

(This highly improper integral is actually quite useful.)

Operators

Linear operators acting on wavefunctions were introduced on p. 38, and an analogous introduction must be made here. Operators in the Dirac notation act on statevectors to produce other statevectors and, to avoid confusion, they will be given a typeface of their own: a, b, c, etc.

As an important example, let us find the operator x which is the observable for position. The only sensible requirement is that x must correspond to the multiplication operator x introduced on p. 39; the link is simply made by saying that

$$\text{if} \quad |\Psi\rangle \leftarrow \psi(x) \quad \text{then} \quad \mathsf{x}|\Psi\rangle \leftarrow x\psi(x).$$

Hence, 'premultiplying' by $\langle x|$,

$$\langle x|\mathsf{x}|\Psi\rangle = x\psi(x) = x\langle x|\Psi\rangle,$$

and since this is true for any state Ψ, we have

$$\langle x|\mathsf{x} = x\langle x|.$$

This may actually be taken as the *definition* of the operator x: it specifies the action of x on each of the vectors $\langle x|$, and therefore on *any* vector $\langle\Psi|$ when expressed in terms of the $\langle x|$. Similarly,

$$\mathsf{x}|x\rangle = |x\rangle x.$$

The last two relations are of eigenvalue-eigenvector form, and show that $\langle x|$ and $|x\rangle$ are *eigenvectors* of x with eigenvalue x: thus they represent the state for which the particle will be found at x with certainty. As discussed on p. 42, such a state cannot be attained in practice; the mathematical expression of this fact is that $|x\rangle$ cannot be normalized:

$$\langle x|x\rangle = \delta(0).$$

In spite of this, they are too useful to be banished from the formalism.

The observable p for the particle *momentum* may be defined in a similar way by specifying its action on each of the vectors $\langle x|$. We write

$$\langle x|\mathsf{p} = -i\hbar \frac{\mathrm{d}}{\mathrm{d}x}\langle x|.$$

To see that this is correct, it is sufficient to note that 'postmultiplying' by $|\Psi\rangle$ gives

$$\mathsf{p}|\Psi\rangle \leftarrow \langle x|\mathsf{p}|\Psi\rangle = -i\hbar \frac{\mathrm{d}}{\mathrm{d}x}\langle x|\Psi\rangle = -i\hbar \frac{\mathrm{d}\psi(x)}{\mathrm{d}x};$$

the link with our previous work is therefore complete.

Definitions of more complicated operators go through in a similar way. For example, the Hamiltonian

$$\mathsf{H} = \frac{1}{2m}\mathsf{p}^2 + V(\mathsf{x})$$

will be completely specified by

$$\langle x|\mathsf{H} = \left(-\frac{\hbar^2}{2m}\frac{\mathrm{d}^2}{\mathrm{d}x^2} + V(x)\right)\langle x|.$$

The **matrix element** of an operator Q between two states Φ and Ψ is the expression $\langle\Phi|\mathsf{Q}|\Psi\rangle$; some authors use the term only when the two states are both eigenstates of the Hamiltonian. An operator Q may be defined by defining enough of its matrix elements: for example, $\langle x|\mathsf{Q}|x'\rangle$ for all x and x'.

The **Hermitian conjugate** K^+ of an operator K may be defined through its matrix elements by

$$\langle\Psi|\mathsf{K}^+|\Phi\rangle = \langle\Phi|\mathsf{K}|\Psi\rangle^*$$

for any states Φ and Ψ. A **Hermitian** operator K satisfies $\mathsf{K} = \mathsf{K}^+$; such operators are sometimes called **real**, since their mean value for any state is always real.

Representations

In ordinary vector algebra, calculations are almost always done in terms of vector components in a Cartesian system of coordinates, and a *wise choice* of axes can be very helpful. The same is true in the space of statevectors: out of a tremendous choice of **representations** of a quantum system, only a very few are ever used with any regularity, namely those whose *basis vectors are eigenvectors of an important observable*. Any other choice is unwise. The important observables are almost all either 'configurational'—coordinates of various kinds—or 'conserved'—momentum, angular momentum, energy and so on.

Incidentally, not every Hermitian operator is an observable. The mean value of a Hermitian operator is always real, and this might be thought to be enough (see p. 18). However, the pleasant eigenvalue behaviour of Hermitian matrices (p. 15) does *not* necessarily carry over to the more general case. An **observable**, in addition to being Hermitian, must have enough eigenvectors to provide a basis for a representation. In practice, the Hamiltonian is always an observable, and so is any coordinate used to describe the configuration of a system.

Let us summarize the work of this chapter so far.

(i) Our chosen system is a point particle moving in one dimension.

(ii) Our chosen representation is the x-representation: the basis states are the eigenvectors $|x\rangle$, labelled for convenience with the relevant eigenvalue x.

(iii) Any state whatever of the system is a linear superposition of the basis states:

$$|\Psi\rangle = \int |x\rangle\, dx \psi(x).$$

(iv) Thus we may choose to do calculations using the 'coefficients' $\psi(x)$ (the wavefunction). Among other things, the Hamiltonian needs to be appropriately translated into this representation:

$$\langle x|\left(\frac{1}{2m}\mathsf{p}^2 + V(\mathsf{x})\right) = \left(-\frac{\hbar^2}{2m}\frac{\partial^2}{\partial x^2} + V(x)\right)\langle x|$$

On the other hand, we may decide *not* to use $\psi(x)$ for calculations, because another representation may make things much simpler. How may we make the change?

As an example, consider the same system, but in the p-representation (often used in scattering problems, in fact). The first step is to find the eigenstates of p; we shall have to do this in the x-representation, as that is all we have so far. The eigenvalue equation

$$\mathsf{p}|p\rangle = |p\rangle p$$

becomes, in the x-representation,

$$-i\hbar\frac{\partial}{\partial x}\langle x|p\rangle = \langle x|p\rangle p,$$

with the solution

$$\langle x|p\rangle = (2\pi\hbar)^{-\frac{1}{2}} \exp(ipx/\hbar)$$

(The disposable multiplying constant has been fixed with future normalization in mind.) This tells us all we need to know to change to the p-representation; an application of the unit-operator trick now gives for *any* state Ψ

$$|\Psi\rangle \leftarrow \langle p|\Psi\rangle = \langle p| \int_{-\infty}^{\infty} |x\rangle\, dx \langle x|\Psi\rangle$$

$$= (2\pi\hbar)^{-\frac{1}{2}} \int_{-\infty}^{\infty} \exp(-ipx/\hbar)\, dx \psi(x)$$

$$= \phi(p), \quad \text{say.}$$

This is the wavefunction in the p-representation. It is an entirely different function from $\psi(x)$ (in fact, it is clearly the Fourier transform of $\psi(x)$); nevertheless it is just as good for calculations as $\psi(x)$, in principle.

There is still some tidying to be done. First, normalization: we have, using the unit operator trick again,

$$\langle p|p'\rangle = \langle p| \int_{-\infty}^{\infty} |x\rangle\, dx \langle x|p'\rangle$$

$$= \frac{1}{2\pi\hbar} \int_{-\infty}^{\infty} \exp(-ipx/\hbar)\, dx \exp(ip'x/\hbar)$$

$$= \frac{1}{2\pi} \int_{-\infty}^{\infty} \exp(-i(p-p')k)\, dk$$

$$= \delta(p-p')$$

(problem 4, p. 116). Then, by the basic property of the δ-function,

$$\int |p\rangle\, dp \langle p|p'\rangle = |p'\rangle$$

for any p', showing that

$$\int_{-\infty}^{\infty} |p\rangle\, dp \langle p| = 1$$

is another way of writing the unit operator. Finally,

$$|\Psi\rangle = \int_{-\infty}^{\infty} |p\rangle\, dp \langle p|\Psi\rangle$$

$$= \int_{-\infty}^{\infty} |p\rangle\, dp \phi(p),$$

showing that the general state $|\Psi\rangle$ may be written as a linear combination of the basis states $|p\rangle$ with the help of the function ϕ, exactly as it may be written as a combination of the entirely different basis states $|x\rangle$, using the entirely different function ψ.

A word of caution: in the expression $|x\rangle$, the symbol x is

(i) a *label* to show that $|x\rangle$ is an eigenvector of x, and
(ii) a *parameter* to show which eigenvector it is.

Thus 'x' carries a double burden of meaning. An expression like $|2\rangle$—if '2' is intended to be a value and not a label—has little meaning: is it an eigenvector of x or, perhaps, of p? We would need to write, for example, $|x = 2\rangle$ or something similar.

Electron spin: the m_z-representation

An electron has a spin of magnitude $\frac{1}{2}\hbar$. The z-component m_z of this spin may be used to provide the m_z-representation, as follows. The operator m_z has two eigenvalues with corresponding eigenvectors (p. 68)

$$\langle +|\mathsf{m}_z = \tfrac{1}{2}\hbar\langle +| \quad \text{and} \quad \langle -|\mathsf{m}_z = -\tfrac{1}{2}\hbar\langle -|;$$

the eigenvectors are arranged to be normalized, besides being orthogonal:

$$\langle +|+\rangle = 1 = \langle -|-\rangle, \langle +|-\rangle = 0.$$

(There is no question here of trouble with δ-functions, as we are not dealing with a continuous range of eigenvalues.) It follows (problem below) that the unit operator may be written

$$1 = |+\rangle\langle +| \; + \; |-\rangle\langle -|.$$

Suppose that $|S\rangle$ is an arbitrary state of the spin. Then the unit-operator trick allows us to write $|S\rangle$ as a linear superposition

$$\begin{aligned} |S\rangle &= |+\rangle\langle +|S\rangle \; + \; |-\rangle\langle -|S\rangle \\ &= |+\rangle c_+ \; + \; |-\rangle c_- \end{aligned}$$

where the two numbers $c_+ = \langle +|S\rangle$ and $c_- = \langle -|S\rangle$ represent S in the m_z-representation: they are the numbers we actually do calculations with. It is usual to arrange them in a column vector

$$c = \begin{pmatrix} c_+ \\ c_- \end{pmatrix} = \begin{pmatrix} \langle +|S\rangle \\ \langle -|S\rangle \end{pmatrix} = \begin{pmatrix} \langle +| \\ \langle -| \end{pmatrix} |S\rangle;$$

the last equality is by a natural convention.

When S is represented by a column vector in this way, then operators are represented by *matrices*; for example

$$\begin{pmatrix} \langle +| \\ \langle -| \end{pmatrix} \mathsf{m}_z = \begin{pmatrix} \langle +|\mathsf{m}_z \\ \langle -|\mathsf{m}_z \end{pmatrix} = \tfrac{1}{2}\begin{pmatrix} \hbar\langle +| \\ -\hbar\langle -| \end{pmatrix} = \tfrac{1}{2}\hbar\begin{pmatrix} 1 & 0 \\ 0 & -1 \end{pmatrix}\begin{pmatrix} \langle +| \\ \langle -| \end{pmatrix}$$

shows that m_z is to be represented by the Pauli matrix $\frac{1}{2}\hbar\sigma_z$. The developments on p. 69 show that m_x and m_y must be represented by the matrices $\frac{1}{2}\hbar\sigma_x$, $\frac{1}{2}\hbar\sigma_y$, in the same kind of way.

All this looks rather different from the corresponding formalism for a particle in one dimension. It must be emphasized that the procedure is exactly the same. The notational differences arise because in the one case it is more convenient to use matrices, in the other, functions.

PROBLEM

Verify that

$$|+\rangle = \{|+\rangle\langle +| + |-\rangle\langle -|\}|+\rangle,$$

along with a similar relation for $|-\rangle$.

The particle in a box

In problem 2 (p. 61) is given the example of a particle confined to a finite region, namely, a one-dimensional box. In the x-representation, the energy eigenstates $|E_n\rangle$ are given by

$$E_n = \frac{\hbar^2\pi^2}{2ma^2}n^2 \quad \text{and} \quad \langle x|E_n\rangle = \left(\frac{2}{a}\right)^{\frac{1}{2}} \sin\frac{n\pi x}{a}.$$

Knowing this, there is nothing to stop us using the H-representation, if we prefer, the basis being the eigenvectors $|E_n\rangle$ of the Hamiltonian H. We therefore have a choice of ways of representing a general state

$$|\Psi\rangle = \int_0^a |x\rangle\, \mathrm{d}x\psi(x) \quad \text{or} \quad \sum_{n=1}^{\infty} |E_n\rangle c_n:$$

the function $\psi(x)$ and the set of coefficients c_n contain exactly equivalent information about the state. Calculations using ψ will *look* very different from those using the cs, but the two will be intimately connected by the Fourier relation

$$c_n = \langle E_n|\Psi\rangle = \int_0^a \langle E_n|x\rangle\, \mathrm{d}x\langle x|\Psi\rangle = \left(\frac{2}{a}\right)^{\frac{1}{2}} \int_0^a \sin\frac{n\pi x}{a}\, \mathrm{d}x\psi(x).$$

When the basics of quantum mechanics were being correctly formulated for the first time in 1925, Werner Heisenberg and Max Born were essentially concerned with why atomic systems make ‘transitions’ between discrete energy levels; because of this emphasis their work was carried on in what we would now call the H-representation, in which states are written as superpositions of levels. It was then natural that the superposition coefficients c_n should be arranged as column vectors, and operators as matrices: the formalism was called **matrix mechanics**. Erwin Schrödinger, on the other hand, started with de Broglie’s view of the electron as a wave extended through space; for him, the role of $\psi(x)$ was to describe the shape of this wave. The natural consequence was a formalism which we would now call the x-representation: **wave mechanics**. The two are so unlike on the surface that Schrödinger and Heisenberg were at first each profoundly suspicious of the other's contribution; in the event it was Schrödinger who first demonstrated that the two approaches are equivalent.

An acute difficulty that faced these pioneers was that they knew little about matrices. Born was profoundly helped by a chance meeting in a railway train with Pascual Jordan, a mathematician with a considerable experience with matrices, who offered his services for the work. Even so, there is little doubt that, had the physicists known what the mathematicians had already known for decades, their passage would have been much easier. For example:

> Hilbert was having a great laugh on Born and Heisenberg and the Göttingen theoretical physicists because when they first discovered matrix mechanics they were having, of course, the same kind of trouble that everybody else had in trying to solve problems and to manipulate and really do things with matrices. So they went to Hilbert for help, and Hilbert said the only times that he had ever anything to do with matrices was when they came up as a sort of by-product of the eigenvalues of the boundary-value problem of a differential equation. So if you look for the differential equation which has these matrices you can probably do more with that. They had thought it was a goofy idea and that Hilbert did not know what he was talking about. So he was having a lot of fun pointing out to them that they could have discovered Schrödinger's wave mechanics six months earlier if they had paid a little more attention to him.
>
> Edward U. Condon. 60 years of quantum physics. *Physics Today* 15, 37–49 (10 October 1962).

PROBLEMS

1. Show that $\sum_{n=1}^{\infty} |E_n\rangle\langle E_n|$ is the unit operator. (Multiply on the right by $|E_m\rangle$ and use $\langle E_n|E_m\rangle = \delta_{mn}$.)

2. Show that $\psi(x) = \left(\dfrac{2}{a}\right)^{\frac{1}{2}} \sum_{n=1}^{\infty} c_n \sin n\pi x/a$.

3. Show that there is no p-representation. (The eigen-problem

$$-i\hbar\frac{\partial\psi}{\partial x} = p\psi, \qquad \psi(0) = 0, \qquad \psi(a) = 0$$

has no solutions.)

The spinless particle in three dimensions: cartesian representations

A particle in three dimensions has a position $\boldsymbol{r} = (x, y, z)$; hence we may set up the xyz-representation, wherein the wavefunction ψ is a function of the cartesian components of $\boldsymbol{r}$:

$$\psi(x, y, z) = \langle xyz|\Psi\rangle, \quad \text{or more briefly,} \quad \psi(\boldsymbol{r}) = \langle \boldsymbol{r}|\Psi\rangle.$$

Thus the basis vectors $\langle xyz|$ are defined; each is a *simultaneous* eigenvector of the *three* observables x, y, z:

$$\langle xyz|\mathsf{x} = x\langle xyz|, \text{ etc.},$$

and the general state may be written

$$|\Psi\rangle = \int |xyz\rangle\, \mathrm{d}x\, \mathrm{d}y\, \mathrm{d}z\, \psi(x, y, z).$$

The normalization for the basis vectors involves a three-dimensional δ-function

$$\langle xyz|x'y'z'\rangle = \delta^3(\boldsymbol{r}-\boldsymbol{r}'),$$

such that

$$\delta^3(\boldsymbol{r}) = 0 \qquad (\boldsymbol{r} \neq 0), \qquad \int \mathrm{d}x\, \mathrm{d}y\, \mathrm{d}z\, \delta^3(\boldsymbol{r}) = 1,$$

where the integral is over any region including the origin. As in one dimension, there is a wealth of 'approximations' to this δ-function, but perhaps the simplest way of writing it is

$$\delta^3(\boldsymbol{r}) = \delta(x)\delta(y)\delta(z).$$

The linear momentum eigenstates $|\boldsymbol{p}\rangle$ (cf. p. 63) are obtained from

$$-\mathrm{i}\hbar\nabla\langle \boldsymbol{r}|\boldsymbol{p}\rangle = \langle \boldsymbol{r}|\boldsymbol{p}\rangle \boldsymbol{p}$$

and may be used as the basis for the **p**-representation,

$$\phi(\boldsymbol{p}) = \langle \boldsymbol{p}|\Psi\rangle.$$

If the states $|\boldsymbol{p}\rangle$ are normalized according to

$$\langle \boldsymbol{p}|\boldsymbol{p}'\rangle = \delta^3(\boldsymbol{p}-\boldsymbol{p}'),$$

then (as on p. 119) ψ and ϕ are related by a Fourier transformation,

$$\phi(\boldsymbol{p}) = (2\pi\hbar)^{-3/2} \int_{\text{all space}} \exp(-\mathrm{i}\boldsymbol{p}.\boldsymbol{r}/\hbar)\, \mathrm{d}^3\boldsymbol{r}\, \psi(\boldsymbol{r}).$$

The spinless particle in three dimensions: polar representations

We may just as well use *polar coordinates* to specify the position of a particle, and this leads to the $r\theta\phi$-representation

$$\psi(r, \theta, \phi) = \langle r\theta\phi|\Psi\rangle$$

without trouble. The volume element in polar coordinates is however $r^2 \sin\theta \, dr \, d\theta \, d\phi$, and it is therefore convenient to normalize the basis so that the unit operator is

$$1 = \int_0^\infty \int_0^\pi \int_0^{2\pi} |r\theta\phi\rangle r^2 \sin\theta \, dr \, d\theta \, d\phi \langle r\theta\phi|;$$

thus we have

$$\langle r\theta\phi|r'\theta'\phi'\rangle = (r^2 \sin\theta)^{-1}\delta(r-r')\delta(\theta-\theta')\delta(\phi-\phi').$$

Apart from this small new feature, everything goes through as before.

There is another, rather mixed, representation which is particularly important for a particle in a spherically symmetric potential; this is the rlm-representation. The basis states are

$$|rlm\rangle : \begin{cases} \text{distance from O is } r \\ \text{(angular momentum about O)}^2 \text{ is } \hbar^2 l(l+1) \\ \text{angular momentum about O}z \text{ is } \hbar m. \end{cases}$$

There are some restrictions: r is a continuous parameter, $0 \leqslant r < \infty$, while l and m are integers, $0 \leqslant l$, $-l \leqslant m \leqslant l$. Because this set of basis states has both continuous and discrete features, the normalization requirement is

$$\langle rlm|r'l'm'\rangle = r^{-2}\delta(r-r')\delta_{ll'}\delta_{mm'},$$

while the general state may be written

$$|\Psi\rangle = \sum_{l=0}^{\infty} \sum_{m=-l}^{l} \int_0^\infty |rlm\rangle r^2 \, dr R_{lm}(r).$$

Of course, we have seen all this before; these developments are intimately connected with the separation of variables carried out in Chapter 8 (pp. 79–82). For example, it may be shown that

$$\langle rlm|\mathbf{p}^2 = -\hbar^2\left(\frac{1}{r}\frac{\partial^2}{\partial r^2}r + \frac{l(l+1)}{r^2}\right)\langle rlm|$$

is the counterpart of

$$\langle xyz|\mathbf{p}^2 = -\hbar^2\nabla^2\langle xyz|,$$

and that

$$\langle rlm|r'\theta\phi\rangle = r^{-2}\delta(r-r')Y_{lm}(\theta\phi).$$

The electron

An electron has position $\boldsymbol{r}$ and spin $\frac{1}{2}\hbar\boldsymbol{\sigma}$, and an adequate representation must therefore be a synthesis of an $\mathbf{r}$-representation and, say, an m_z-

representation. The synthesis is the **direct product** (or **tensor** or **Kronecker** product) of the two representations, immediately obtained by setting up the rm_z-representation, with the basis states $\langle \boldsymbol{r} m_z|$. Since m_z can have only two values $\pm\frac{1}{2}\hbar$, it is convenient to think in terms of *two* wavefunctions

$$\psi_+(\boldsymbol{r}) = \langle \boldsymbol{r}, \tfrac{1}{2}\hbar|\Psi\rangle \quad \text{and} \quad \psi_-(\boldsymbol{r}) = \langle \boldsymbol{r}, -\tfrac{1}{2}\hbar|\Psi\rangle,$$

arranged as a two-component column vector, which is at the same time a function of position,

$$\psi(\boldsymbol{r}) = \begin{pmatrix} \psi_+(\boldsymbol{r}) \\ \psi_-(\boldsymbol{r}) \end{pmatrix},$$

as already described (p. 73). Operators whose source is the r-representation (like $\boldsymbol{r}, \nabla$) act on $\psi(\boldsymbol{r})$ as a function of $\boldsymbol{r}$, while operators from the m_z-representation (like $\boldsymbol{\sigma}$) act on $\psi(\boldsymbol{r})$ as a two-component vector.

Direct products

Such procedures for a direct product are quite general. If we have an A-representation for system 1, and a B-representation for system 2, then the quantum mechanics of the combined system (1 with 2) may be described in terms of the AB-representation. The action of any operator Q_1 belonging to system 1 must be extended in a very obvious way: if originally, for example,

$$\langle a|\mathsf{Q}_1 = \sum_{a'} q_{aa'}\langle a'|$$

then

$$\langle ab|\mathsf{Q}_1 = \sum_{a'} q_{aa'}\langle a'b|.$$

(This is the matrix case in particular: the general rule is that the inclusion of b must make no difference.)

From time to time we shall need expressions like

$$M = \langle ab|\mathsf{Q}_1\mathsf{Q}_2|a'b'\rangle$$

where $\mathsf{Q}_1, \mathsf{Q}_2$ are operators belonging to systems 1 and 2 respectively. (This is the most usual way to set up an interaction between two systems.) The rule is that M has two factors, one evaluated in system 1 and the other in system 2:

$$M = \langle a|\mathsf{Q}_1|a'\rangle\langle b|\mathsf{Q}_2|b'\rangle.$$

Time-dependence

Throughout this rather long chapter we have ignored the time t. The wavefunction $\psi(xt)$—or whatever the representation happens to be—is expected to change as time passes. Also,

$$\psi(xt) = \langle x|\Psi\rangle,$$

and we must ask: which factor changes with time, $\langle x|$ or $|\Psi\rangle$? The answer is: either. We have two options,

the **Schrödinger picture:** $\langle x|$ is constant;

the **Heisenberg picture:** $|\Psi\rangle$ is constant.

Both pictures have virtues to commend them.

Throughout this book we keep to the *Schrödinger picture*; that is, the basis of any representation does not change, while the vector representing the state varies according to the Hamiltonian equation

$$i\hbar \frac{\mathrm{d}}{\mathrm{d}t}|\Psi t\rangle = \mathsf{H}|\Psi t\rangle.$$

The Schrödinger equation in any particular representation is easily obtained. For example, if

$$\mathsf{H} = \frac{1}{2m}\mathsf{p}^2 + V(\mathsf{x}), \qquad \text{i.e. } \langle x|\mathsf{H} = \left(-\frac{\hbar^2}{2m}\frac{\partial^2}{\partial x^2} + V(x)\right)\langle x|$$

then

$$\begin{aligned} i\hbar \frac{\partial \psi(xt)}{\partial t} &= \langle x|i\hbar \frac{\mathrm{d}}{\mathrm{d}t}|\Psi t\rangle \\ &= \langle x|\mathsf{H}|\Psi t\rangle \\ &= \left(-\frac{\hbar^2}{2m}\frac{\partial^2}{\partial x^2} + V(x)\right)\psi(xt), \end{aligned}$$

as expected.

11. Harmonic motion

The importance of harmonic oscillation

HARMONIC motion is perhaps the most important of all dynamical motions, whether in classical or quantum mechanics. The enormous progress over four centuries in the accuracy of timekeeping may well have begun in Pisa Cathedral in 1581 with Galileo's discovery of the most characteristic property of harmonic motion: he noticed that the period of swing of a large chandelier was independent of its amplitude. (We are sometimes told that Galileo used his own pulse as a 'clock', but not what the excitement of discovery may have done to his heart rate!) Harmonic motion is central for any periodic system having *normal modes* (at least approximately); two examples of interest to us are the internal vibrations of molecules, and the electromagnetic field. Apart from all these practical matters, harmonic motion provides one of the best illustrations of the methods of quantum mechanics. It therefore deserves a chapter of its own.

The **simple harmonic oscillator** executes harmonic motion, and has just one coordinate x. An example of such an oscillator has been given on p. 48; the Hamiltonian is

$$H = -\frac{\hbar^2}{2m}\frac{\partial^2}{\partial x^2} + \tfrac{1}{2}m\omega^2 x^2$$

(using $k = m\omega^2$, where ω is the angular frequency of the oscillator). The Schrödinger equation is

$$i\hbar\frac{\partial\psi}{\partial t} = -\frac{\hbar^2}{2m}\frac{\partial^2\psi}{\partial x^2} + \tfrac{1}{2}m\omega^2 x^2\psi.$$

From the outset, we shall 'scale' the oscillator to a standard form: set

$$s = x/x_0 \quad \text{with} \quad x_0^2 = \hbar/m\omega;$$

that is, the displacement will be measured in multiples of the length x_0. In terms of s, the Schrödinger equation for the wavefunction $\psi(s, t)$ is

$$i\hbar\frac{\partial\psi}{\partial t} = \hbar\omega\left(-\tfrac{1}{2}\frac{\partial^2\psi}{\partial s^2} + \tfrac{1}{2}s^2\psi\right).$$

Even the $\hbar$ may now be cancelled, and there remains only one parameter ω to define the oscillator. Any simple harmonic oscillator may be 'scaled' to this form; we shall find this useful later.

Stationary states: the direct attack

The eigenfunctions for stationary states satisfy

$$E\psi = \hbar\omega\left(-\tfrac{1}{2}\frac{\partial^2\psi}{\partial s^2} + \tfrac{1}{2}s^2\psi\right).$$

We have already seen that the spectrum is *discrete* (p. 48), and that the first few eigenfunctions all contain a factor $\exp(-\tfrac{1}{2}s^2)$, the other factor being a polynomial. Therefore we hope to make progress by setting

$$\psi = f(s)\exp(-\tfrac{1}{2}s^2),$$

and, in consequence, (with $f' \equiv \partial f/\partial s$, etc.)

$$\psi'' = (f'' - 2sf' + (s^2-1)f)\exp(-\tfrac{1}{2}s^2).$$

Inserting in the original equation now gives the equation for f,

$$Ef = \hbar\omega(-\tfrac{1}{2}f'' + sf' + \tfrac{1}{2}f).$$

This is one of those problems which is amenable to the power series approach. None of the terms is singular for any value of s, so we may expand f as a Taylor series, convergent for all s,

$$f(s) = \sum_{j=0}^{\infty} c_j s^j,$$

and substitute in the equation. Collecting powers of s^j in the result gives

$$Ec_j = \hbar\omega\{-\tfrac{1}{2}(j+2)(j+1)c_{j+2} + sc_j + \tfrac{1}{2}c_j\} \qquad \text{for } j \geqslant 0,$$

that is,

$$\tfrac{1}{2}\hbar\omega(j+2)(j+1)c_{j+2} = \{(j+\tfrac{1}{2})\hbar\omega - E\}c_j \qquad (j \geqslant 0).$$

If we know c_0 and c_1, any subsequent coefficient may be obtained by applying this relation a sufficient number of times.

It remains to ensure that $\psi(s) \to 0$ as $|s| \to \infty$. Now, for large j (unless the coefficients are accidentally zero beyond a certain point),

$$\frac{c_{j+2}}{c_j} = \frac{2}{j} + 0(1/j^2);$$

that is, the coefficients show a behaviour similar to those in the Taylor series for $\exp s^2$. Consequently, $\psi(\equiv f\exp(-\tfrac{1}{2}s^2))$ is inevitably unbounded as s goes to infinity in either direction, and this will not do. The series for f must therefore *terminate*. This happens only if—compare with with the similar discussion on p. 96—

$$E = E_n \equiv (n+\tfrac{1}{2})\hbar\omega \qquad (n \text{ some integer } \geqslant 0),$$

when the final term in the series will be $c_n s^n$. Since the relation between the coefficients connects *alternate* terms, the resulting polynomial will be *even* if n is even, and *odd* if n is odd.

The first few (unnormalized) functions are, including their time dependence,

$$\psi_0 = \exp(-\tfrac{1}{2}s^2)\exp(-\tfrac{1}{2}\mathrm{i}\omega t) \qquad (E = \tfrac{1}{2}\hbar\omega)$$

$$\psi_1 = s\exp(-\tfrac{1}{2}s^2)\exp(-\tfrac{3}{2}\mathrm{i}\omega t) \qquad (E = \tfrac{3}{2}\hbar\omega)$$

$$\psi_2 = (2s^2-1)\exp(-\tfrac{1}{2}s^2)\exp(-\tfrac{5}{2}\mathrm{i}\omega t) \qquad (E = \tfrac{5}{2}\hbar\omega)$$

etc.

The algebraic formulation of the simple harmonic oscillator

We shall carry out the solution of the simple harmonic oscillator again, this time using an algebraic formalism based on the Dirac notation. This approach has several definite advantages in problems of a certain special type, of which the harmonic oscillator is one.

The starting-point is the fact that the Hamiltonian is almost factorizable as the difference of two squares,

$$\left(s-\frac{\mathrm{d}}{\mathrm{d}s}\right)\left(s+\frac{\mathrm{d}}{\mathrm{d}s}\right)\psi(s) = \left(s^2-\frac{\mathrm{d}^2}{\mathrm{d}s^2}-1\right)\psi(s),$$

the extra -1 arising from the action of $\mathrm{d}/\mathrm{d}s$ in the first factor on the s in the second. We shall introduce the s-representation by writing

$$\psi(s) = \langle s|\psi\rangle,$$

and define two operators a, a^+ by the relations (using the technique of p. 116)

$$\langle s|\mathsf{a} = 2^{-\frac{1}{2}}\left(s+\frac{\mathrm{d}}{\mathrm{d}s}\right)\langle s|,$$

$$\langle s|\mathsf{a}^+ = 2^{-\frac{1}{2}}\left(s-\frac{\mathrm{d}}{\mathrm{d}s}\right)\langle s|.$$

(As the notation suggests, a^+ is in fact the Hermitian conjugate of a.) Then

$$\hbar\omega\langle s|\mathsf{a}^+\mathsf{a} = \frac{\hbar\omega}{2}\left(s-\frac{\mathrm{d}}{\mathrm{d}s}\right)\left(s+\frac{\mathrm{d}}{\mathrm{d}s}\right)\langle s|$$

$$= \frac{\hbar\omega}{2}\left(s^2-\frac{\mathrm{d}^2}{\mathrm{d}s^2}-1\right)\langle s| = \langle s|(\mathsf{H}-\tfrac{1}{2}\hbar\omega),$$

whence

$$\mathsf{H} = \hbar\omega(\mathsf{a}^+\mathsf{a}+\tfrac{1}{2}) = \hbar\omega(\mathsf{n}+\tfrac{1}{2}),$$

where, for convenience, we have introduced

$$\mathsf{n} = \mathsf{a}^+\mathsf{a}.$$

(In view of the last section's work, we will not be surprised when we find that the eigenvalues of n are the non-negative integers.)

Similarly,

$$\langle s|(\mathsf{aa}^+ - \mathsf{a}^+\mathsf{a}) = \tfrac{1}{2}\left\{\left(s+\frac{\mathrm{d}}{\mathrm{d}s}\right)\left(s-\frac{\mathrm{d}}{\mathrm{d}s}\right)-\left(s-\frac{\mathrm{d}}{\mathrm{d}s}\right)\left(s+\frac{\mathrm{d}}{\mathrm{d}s}\right)\right\}\langle s| = \langle s|,$$

from which we deduce

$$[\mathsf{a}, \mathsf{a}^+] \equiv \mathsf{aa}^+ - \mathsf{a}^+\mathsf{a} = 1.$$

Stationary states: the algebraic attack

It is a remarkable fact that the two relations

$$\mathsf{n} = \mathsf{a}^+\mathsf{a} \quad \text{and} \quad \mathsf{aa}^+ - \mathsf{a}^+\mathsf{a} = 1$$

are all we need to deal with the simple harmonic oscillator completely. In a very effective way, we have extracted the *essential structure* of the problem. Finding the stationary states reduces to finding the eigenvalues and eigenvectors of n.

First, *none of the eigenvalues of* n *is negative*. For, if $|n\rangle$ is an eigenvector with eigenvalue n,

$$n\langle n|n\rangle = \langle n|\mathsf{n}|n\rangle = \langle n|\mathsf{a}^+ . \mathsf{a}|n\rangle,$$

and the right side is *not negative*, being the scalar product of the vector $\mathsf{a}|n\rangle$ with its own hermitian conjugate. Also, $\langle n|n\rangle > 0$, for a similar reason. Hence the eigenvalue $n \geqslant 0$.

We are now ready for the main attack. Consider

$$\begin{aligned} \mathsf{an} - \mathsf{na} &= \mathsf{aa}^+\mathsf{a} - \mathsf{a}^+\mathsf{aa} \\ &= (\mathsf{aa}^+ - \mathsf{a}^+\mathsf{a})\mathsf{a} \\ &= \mathsf{a} \end{aligned}$$

or

$$\mathsf{na} = \mathsf{a}(\mathsf{n} - 1).$$

This result gives us a hold on the eigenvectors of n in the following way. Suppose $|n\rangle$ is any eigenvector of n (with eigenvalue $n \geqslant 0$, of course):

$$\mathsf{n}|n\rangle = |n\rangle n.$$

Then

$$\mathsf{n}\mathsf{a}|n\rangle = \mathsf{a}(\mathsf{n}-1)|n\rangle \qquad \text{(by our last result)}$$
$$= \mathsf{a}|n\rangle(n-1).$$

Hence, *provided* $\mathsf{a}|n\rangle \neq 0$, we have found *another* eigenvector $\mathsf{a}|n\rangle$, this time with eigenvalue $n-1$.

We may repeat this process to obtain a sequence of eigenvectors $|n\rangle$, $\mathsf{a}|n\rangle$, $\mathsf{a}^2|n\rangle, \ldots,$ with steadily decreasing eigenvalues $n, n-1, n-2, \ldots$. This sequence must stop before the eigenvalue goes negative; the only way in which the sequence may stop is for $\mathsf{a}|m\rangle$ to become zero, where $|m\rangle$ is the last eigenvector in the sequence. But then

$$\mathsf{n}|m\rangle = \mathsf{a}^+\mathsf{a}|m\rangle = 0,$$

showing that $|m\rangle$ is an eigenvector with eigenvalue 0. Thus the sequence of decreasing eigenvalues ends with 0; consequently, they are *consecutive non-negative integers*.

To see that the eigenvalues of n are *all* the non-negative integers, we must move in the *other* direction. By a similar technique, we show that

$$\mathsf{n}\mathsf{a}^+ = \mathsf{a}^+(\mathsf{n}+1)$$

and

$$\mathsf{n}\mathsf{a}^+|n\rangle = \mathsf{a}^+|n\rangle(n+1);$$

that is, $\mathsf{a}^+|n\rangle$ is another eigenvector with eigenvalue $n+1$ (unless $\mathsf{a}^+|n\rangle$ happens to be zero; but, since

$$\langle n|\mathsf{a}\mathsf{a}^+|n\rangle = \langle n|(\mathsf{a}^+\mathsf{a}+1)|n\rangle = \langle n|(\mathsf{n}+1)|n\rangle = \langle n|n\rangle(n+1) > 0,$$

this can never happen.)

From all this it follows that, starting with $|0\rangle$, we may obtain the complete set of eigenvectors of n by repeatedly applying the operator a^+: the eigenvector which corresponds to the eigenvalue n is $(\mathsf{a}^+)^n|0\rangle$.

We ought to consider normalization. Suppose $|n\rangle$ is a *normalized* eigenvector. Then $\mathsf{a}^+|n\rangle$ need not be normalized; in fact,

$$\langle n|\mathsf{a}\,.\,\mathsf{a}^+|n\rangle = \langle n|(\mathsf{n}+1)|n\rangle = (n+1)\langle n|n\rangle = n+1.$$

Thus, the normalized

$$|n+1\rangle = \mathsf{a}^+|n\rangle(n+1)^{-\frac{1}{2}}.$$

If $|0\rangle$ is normalized, we find by induction that

$$|n\rangle = (\mathsf{a}^+)^n|0\rangle(n!)^{-\frac{1}{2}}$$

is also normalized. This is the complete solution to the problem.

The operators a, a^+ are sometimes called **ladder operators**, from the manner in which their application raises or lowers the eigenvalue of n (and hence the energy eigenvalue) step by step.

PROBLEMS

1. The statevector $|\lambda)$ is defined by

$$|\lambda) = \sum_{n=0}^{\infty} |n\rangle \lambda^n (n!)^{-\frac{1}{2}}.$$

(The notation $|\lambda)$ is used to prevent confusion with $|n\rangle$.) Show that $|\lambda)$ is a *right-eigenvector* of a, and $(\lambda|$ is a *left-eigenvector* of a^+, that is,

$$\mathsf{a}|\lambda) = |\lambda)\lambda \quad \text{and} \quad (\lambda|\mathsf{a}^+ = \lambda(\lambda|$$

for any real or complex value of λ whatever.

(The operator a is very lopsided: it has a right-eigenvector for every conceivable eigenvalue, but it has no left-eigenvectors at all. Corresponding remarks may be made about a^+. Of course, a and a^+ are not Hermitian and not observables—though combinations like $\mathsf{s} = 2^{-\frac{1}{2}}(\mathsf{a}+\mathsf{a}^+)$ and $\mathsf{n} = \mathsf{a}^+\mathsf{a}$ are—, so we need not be surprised at their odd properties.)

2. Show that

$$(\mu|\lambda) = \mathrm{e}^{\mu^*\lambda},$$

so that no pair of $|\lambda), |\mu)$ are ever mutually orthogonal. (Substitute the series for $(\mu|$ and $|\lambda)$ in the left side, and use $\langle m|n\rangle = \delta_{mn}$.)

3. Show that in the n-representation the only non-zero matrix elements of a and a^+ are

$$\langle n|\mathsf{a}|n+1\rangle = (n+1)^{\frac{1}{2}} \quad \text{and} \quad \langle n+1|\mathsf{a}^+|n\rangle = (n+1)^{\frac{1}{2}}.$$

(In particular, $\langle 0|\mathsf{a}|1\rangle = 1 = \langle 1|\mathsf{a}^+|0\rangle$.)

Links with the s-representation

There is far more in the work of the last section than appears at first sight. To see how much has really been achieved, let us look at the results in terms of the s-representation. First, the eigenvector $|0\rangle$ satisfies $\mathsf{a}|0\rangle = 0$. In the s-representation, this is

$$\langle s|\mathsf{a}|0\rangle = 2^{-\frac{1}{2}}\left(s + \frac{\mathrm{d}}{\mathrm{d}s}\right)\langle s|0\rangle = 0.$$

This differential equation is easily solved to give

$$\langle s|0\rangle = \pi^{-\frac{1}{4}} \exp\left(-\tfrac{1}{2}s^2\right)$$

(where $\pi^{-\frac{1}{4}}$ is included to give a normalized solution.) This is the wavefunction for $n = 0$, i.e. $E = \frac{1}{2}\hbar\omega$.

All the other wavefunctions are obtained by repeated application of a^+, in the following manner

$$\langle s|1\rangle = \langle s|\mathrm{a}^+|0\rangle = 2^{-\frac{1}{2}}\left(s-\frac{\mathrm{d}}{\mathrm{d}s}\right)\langle s|0\rangle = \pi^{-\frac{1}{4}}2^{-\frac{1}{2}}2s\,.\exp\left(-\tfrac{1}{2}s^2\right)$$

$$\langle s|2\rangle = 2^{-\frac{1}{2}}\langle s|\mathrm{a}^+|1\rangle = 2^{-\frac{1}{2}}\left(s-\frac{\mathrm{d}}{\mathrm{d}s}\right)2^{-\frac{1}{2}}\langle s|1\rangle$$

$$= \pi^{-\frac{1}{4}}2^{-\frac{1}{2}}2^{-1}(4s^2-2)\exp\left(-\tfrac{1}{2}s^2\right)$$

$$\langle s|3\rangle = 3^{-\frac{1}{2}}\langle s|\mathrm{a}^+|2\rangle = \pi^{-\frac{1}{4}}6^{-\frac{1}{2}}2^{-\frac{3}{2}}(8s^3-12s)\exp\left(-\tfrac{1}{2}s^2\right)$$

etc.

In general, the nth wavefunction $\psi_n(s)$, with eigenvalue $E = \hbar\omega(n+\frac{1}{2})$, is

$$\psi_n(s) = \langle s|n\rangle = \pi^{-\frac{1}{4}}(n!)^{-\frac{1}{2}}2^{-\frac{1}{2}n}H_n(s)\exp\left(-\tfrac{1}{2}s^2\right)$$

where $H_n(s)$ is the nth **Hermite polynomial**. Thus we have recovered the solutions given earlier; in addition, their normalization is now correctly given, and we have a simple recursive method of obtaining the nth wavefunction in the sequence. The first three are shown in the figure.

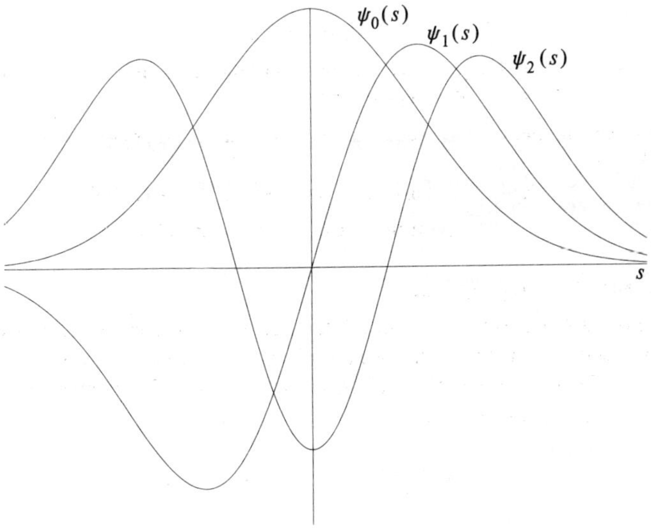

Even the boundary condition $\psi(s) \to 0$ as $s \to \pm\infty$ is taken into account, though it is rather well hidden. For example, we have seen that $\mathrm{a}^+|\psi\rangle$ is never zero. In the s-representation,

$$\langle s|\mathrm{a}^+|\psi\rangle = 2^{-\frac{1}{2}}\left(s-\frac{\mathrm{d}}{\mathrm{d}s}\right)\langle s|\psi\rangle = 0$$

would imply (by straightforward solution)

$$\langle s|\psi\rangle = \text{const. exp}\,(+\tfrac{1}{2}s^2),$$

and this is to be ruled out by the boundary condition. It is *automatically* ruled out in the second approach.

PROBLEMS

1. Show that the relations

$$\psi_n(s) = (2n)^{-\frac{1}{2}}\left(s-\frac{\mathrm{d}}{\mathrm{d}s}\right)\psi_{n-1}(s) \quad \text{and} \quad \psi_{n-1}(s) = (2n)^{-\frac{1}{2}}\left(s+\frac{\mathrm{d}}{\mathrm{d}s}\right)\psi_n(s)$$

imply that

$$H_n = \left(2s-\frac{\mathrm{d}}{\mathrm{d}s}\right)H_{n-1} \quad \text{and} \quad H_{n-1} = \frac{1}{2n}\frac{\mathrm{d}H_n}{\mathrm{d}s}.$$

2. Use

$$H_0 = 1, \qquad H_n = \left(2s-\frac{\mathrm{d}}{\mathrm{d}s}\right)H_{n-1}$$

repeatedly to find H_0, H_1, $H_2 \ldots H_{10}$ in turn. (Check:
$H_{10} = 1024s^{10} - 23\,040s^8 + 161\,280s^6 - 403\,200s^4 + 302\,400s^2 - 30\,240$.)

3. Eliminate H_{n-1} from the two relations of problem 1 to obtain the differential equation

$$\frac{\mathrm{d}^2H_n}{\mathrm{d}s^2} - 2s\frac{\mathrm{d}H_n}{\mathrm{d}s} + 2nH_n = 0.$$

Verify that the first few Hermite polynomials satisfy this equation.

4. Eliminate $\mathrm{d}H_n/\mathrm{d}s$ from the two relations of problem 1 to obtain the relation between three consecutive Hermite polynomials,

$$H_{n+1} - 2sH_n + 2nH_{n-1} = 0.$$

5. In the s-representation, $\mathsf{a}|\lambda) = |\lambda)\lambda$ (problem, p. 132) becomes

$$2^{-\frac{1}{2}}\left(s+\frac{\mathrm{d}}{\mathrm{d}s}\right)\langle s|\lambda) = \langle s|\lambda)\lambda.$$

Show that, taken with $(\lambda|\lambda) = \mathrm{e}^{|\lambda|^2}$, the solution of this equation is

$$\langle s|\lambda) = \mathrm{e}^{\frac{1}{2}\lambda^2}\pi^{-\frac{1}{4}}\exp\,(-\tfrac{1}{2}(s-\lambda\sqrt{2})^2),$$

up to a multiplicative phase factor. Verify also that

$$\langle s|\lambda) = \sum_{n=0}^{\infty} \psi_n(s)\lambda^n(n!)^{-\frac{1}{2}},$$

and hence show that—up to a multiplicative phase factor—

$$\exp\,(2ws - w^2) = \sum_{n=0}^{\infty} \frac{w^n}{n!}H_n(s).$$

(As it happens, the phase factor is 1, and we have a **generating function** for the Hermite polynomials. Such generating functions exist for very many of the collections of polynomials and other functions which appear in theoretical physics, and often provide flexible and succinct routes to important theorems. However, we make no use of them in this book.)

6. The general state $|\Psi\rangle$ may be specified in either the s- or the n-representation

$$|\Psi\rangle = \int_{-\infty}^{\infty} |s\rangle\, \mathrm{d}s\psi(s) \quad \text{or} \quad \sum_{n=0}^{\infty} |n\rangle c_n.$$

Show that

$$c_n = \pi^{-\frac{1}{4}}(n!)^{-\frac{1}{2}}2^{-\frac{1}{2}n} \int_{-\infty}^{\infty} H_n(s) \exp(-\tfrac{1}{2}s^2)\psi(s)\, \mathrm{d}s,$$

and express $\psi(s)$ in terms of the cs. (See p. 121 for a comparable situation.)

The almost rigid diatomic molecule

In classical mechanics it may happen that the motion of a particle is governed by a potential energy with a deep minimum at some point. The motion of the particle is then *approximately* harmonic, provided it remains in the immediate vicinity of the minimum.

The same kind of thing happens in quantum mechanics. As an example, we shall consider the motion of the two atoms (each treated as a unit, with neglect of electronic motion) composing a diatomic molecule. The potential energy is a function $V(r)$ of the separation r, and is *attractive* for large r, but *repulsive* for small r. A typical form of $V(r)$ is given in the figure. The work

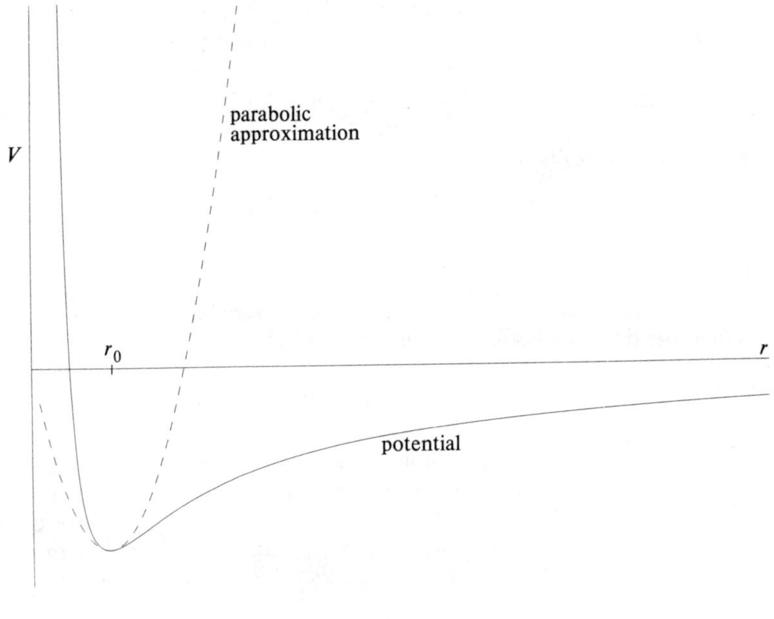

of chapter 9 applies to this case; in particular, the Schrödinger equation is separable, and the radial equation is as on p. 83, with reduced mass μ,

$$-\frac{\hbar^2}{2\mu}\frac{d^2\chi}{dr^2}+v_l(r)\chi = E\chi$$

where the function

$$v_l(r) = V(r)+\frac{\hbar^2 l(l+1)}{2\mu r^2}$$

may also be expected to have a minimum for at least some values of l (this is obviously true if $l = 0$). The method is to expand $v_l(r)$ as a power series about its minimum, and to retain terms only as far as second order, thus approximating the function in the neighbourhood of its minimum by a parabola.

To fix ideas, we shall take

$$V(r) = g\left(\frac{r_0}{2r^2}-\frac{1}{r}\right)$$

where g and r_0 are positive constants. It is straightforward to show that $V(r)$ has its minimum at $r = r_0$, and that

$$V(r_0+\eta) = -\frac{g}{2r_0}+\frac{g}{2r_0^3}\eta^2+0(\eta^3).$$

More generally,

$$v_l(r_l+\eta) = -\frac{g}{2r_l}+\frac{g}{2r_l^3}\eta^2+0(\eta^3),$$

where now the position of equilibrium is

$$r_l = r_0+\frac{\hbar^2 l(l+1)}{\mu g},$$

and $r = r_l+\eta$.

It is convenient to make the simple change of variable from r to η. The radial equation then becomes (with neglect of η^3)

$$-\frac{\hbar^2}{2\mu}\frac{d^2\chi}{d\eta^2}+\frac{g}{2r_l^3}\eta^2\chi = \left(E+\frac{g}{2r_l}\right)\chi.$$

This is of the form for simple harmonic motion with force constant $(\mu\omega^2) = g/r_l^3$. We may therefore write immediately

$$E+\frac{g}{2r_l} = \hbar(n'+\tfrac{1}{2})(g/\mu r_l^3)^{\frac{1}{2}},$$

giving as the approximate eigenvalues for our model of a diatomic molecule

$$E_{n'l} = -\frac{g}{2r_l} + \hbar(n' + \tfrac{1}{2})(g/\mu r_l^3)^{\frac{1}{2}},$$

depending on the two quantum numbers $l = 0, 1, 2, \ldots$ and $n' = 0, 1, 2, \ldots$.

It is instructive to examine these eigenvalues when g is large, that is, when the potential minimum is deep. In this case, the constituent atoms are strongly restricted to having a separation in the neighbourhood of r_0: the molecule is *almost rigid*. We replace r_l in the denominators in $E_{n'l}$ by $r_0 + \hbar^2 l(l+1)/\mu g$ and expand in descending powers of g, giving

$$E_{n'l} = -\frac{g}{2r_0} + \hbar(n' + \tfrac{1}{2})\left(\frac{g}{\mu r_0^3}\right)^{\frac{1}{2}} + \frac{\hbar^2 l(l+1)}{2\mu r_0^2} - \hbar(n' + \tfrac{1}{2})\frac{3\hbar^2 l(l+1)}{2(g\mu^3 r_0^5)^{\frac{1}{2}}} + \ldots.$$

There is no point in going any further, since the terms are already of the same order as terms neglected in the expansion of $v_l(r)$.

The first term in this expression for $E_{n'l}$ is a constant, and represents an unimportant overall energy shift. The second term (of order $g^{\frac{1}{2}}$) gives the energy levels of the approximately harmonic radial vibrations of the molecule. The third term (independent of g) is the kinetic energy of rotation of the molecule, with angular momentum $\hbar\{l(l+1)\}^{\frac{1}{2}}$ and moment of inertia μr_0^2. The fourth term is a correction to the oscillator energy levels, arising from the small change in the effective force constant entailed by the centrifugal stretching of the molecule under rotation (that it is a rotational effect is clear from the way it depends on l).

The effects of these terms are easily observed in the spectra of molecules like HCl, which possess a permanent electric dipole moment. (The rotating dipole moment is necessary to provide a mechanism for electromagnetic radiation.) The energy spectrum of such a molecule is dominated by the *electronic* energy levels E_N, which we have not considered in this section. The possibility of vibration and rotation of the molecule as a whole gives each of these levels a structure which to a good approximation is described by $E_N + E_{n'l}$. For any fixed value of N, varying n' gives a set of levels whose separation is typically about 1/10 of the separation between electronic levels. For any fixed choice of N and n', varying l gives a fine structure, with separations about 1/1000 of the separation between electronic levels. The disparity in the sizes of these separations makes the experimental identification of these effects straightforward.

PROBLEM

The radial equation for the potential

$$V = g\left(\frac{r_0}{2r^2} - \frac{1}{r}\right)$$

is

$$-\frac{\hbar^2}{2\mu}\frac{d^2\chi}{dr^2}+\left\{\frac{gr_0}{2r^2}-\frac{g}{r}+\frac{\hbar^2 l(l+1)}{2\mu r^2}\right\}\chi = E\chi.$$

Apply the general results on pp. 95–97 to obtain the *exact* eigenvalues $E_{n'l} = -\hbar^2\lambda^2/2\mu$, with

$$\lambda\{n'+\tfrac{1}{2}+(\tfrac{1}{4}+l(l+1)+\mu g r_0/\hbar^2)^{\frac{1}{2}}\} = \mu g/\hbar^2.$$

Normal modes

There is an important generalization of the classical simple harmonic oscillator to systems whose configurations require several coordinates x_1, $x_2, \ldots x_N$ for their description. Suppose the Newton's equations of motion are *linear* and *homogeneous* in the *x*s, that is, the equation for the acceleration $\ddot{x}_k$ is

$$\ddot{x}_k = \sum_{j=1}^{N} K_{kj}x_j = K_{k1}x_1 + K_{k2}x_2 + \ldots,$$

there being N such equations. Then it is sufficient to search for the **normal modes** of the system; these are motions of a special kind for which every coordinate executes simple harmonic motion with the *same* angular frequency for each:

$$\ddot{x}_k = -\omega^2 x_k \qquad \text{(each } k\text{)}.$$

Note however that the *amplitude* of oscillation is generally *different* for each coordinate. The possible values of ω, and the ratios of the amplitudes for each such value, are obtained by a simultaneous solution of the N equations (really the eigenvalue–eigenvector equations for the matrix K):

$$-\omega^2 x_k = \sum_{j=1}^{N} K_{kj}x_j.$$

The solution must be *non-trivial*: at least one of the *x*s must be non-zero. The *general* solution for the system is then a linear combination of the normal modes with disposable coefficients. It is natural to regard the entire system as a collection of N simple harmonic oscillators, each of which makes its own independent contribution (with its own characteristic frequency) to the overall motion.

The ideas are best illustrated by a simple example. The three atoms of a (one-dimensional) molecule have equal masses m, and are connected by 'springs' with force constant K. (These springs represent the inter-atomic forces which, for small displacements from equilibrium, may be taken to be harmonic.) The displacements from equilibrium are x, y, z respectively. The

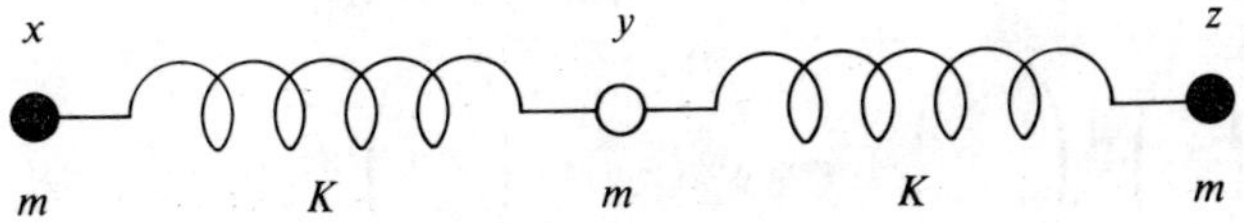

Newtonian equations of motion, one for each of the three particles, are

$$
\begin{aligned}
m\ddot{x} &= -K(x-y),\\
m\ddot{y} &= -K(y-x)-K(y-z),\\
m\ddot{z} &= \qquad\qquad\quad -K(z-y),
\end{aligned}
$$

or, after a slight rearrangement, and including the requirement for a normal mode,

$$
-\omega^2\begin{pmatrix} x\\ y\\ z\end{pmatrix} = \frac{d^2}{dt^2}\begin{pmatrix} x\\ y\\ z\end{pmatrix} = \begin{pmatrix} -1 & 1 & 0\\ 1 & -2 & 1\\ 0 & 1 & -1\end{pmatrix}\begin{pmatrix} x\\ y\\ z\end{pmatrix}.
$$

The matrix has three eigenvectors which happen to be very easy to identify on account of the symmetry of the molecule; they are, along with the corresponding values of ω^2,

$$
\begin{pmatrix} 1\\ 0\\ -1\end{pmatrix}, \text{ for } \omega_1^2 = K/m,
$$

$$
\begin{pmatrix} 1\\ -2\\ 1\end{pmatrix}, \text{ for } \omega_2^2 = 3K/m, \quad \text{and}
$$

$$
\begin{pmatrix} 1\\ 1\\ 1\end{pmatrix}, \text{ for } \omega_3^2 = 0.
$$

The last of these, with $\omega_3^2 = 0$, represents a steady drift of the molecule as a whole, and is of no interest to us here. The general vibrational solution of the problem is thus obtained by superposing the other two normal modes with arbitrary amplitudes A_1, A_2 and phases η_1, η_2,

$$\begin{pmatrix} x \\ y \\ z \end{pmatrix} = \begin{pmatrix} 1 \\ 0 \\ -1 \end{pmatrix} A_1 \cos(\omega_1 t + \eta_1) + \begin{pmatrix} 1 \\ -2 \\ 1 \end{pmatrix} A_2 \cos(\omega^2 t + \eta_2).$$

In general, each of the original coordinates will include a contribution from each of the normal modes, though it happens in this example that y includes no contribution from the first mode, on account of the symmetry of the molecule.

A recipe for quantum normal modes

In general, as we have already mentioned (p. 92), the quantum problem of a system with more than a very few coordinates, $\boldsymbol{r}_1$, $\boldsymbol{r}_2, \ldots$, and Hamiltonian

$$-\frac{\hbar^2}{2m_1}\nabla_1^2 - \frac{\hbar^2}{2m_2}\nabla_2^2 - \ldots + V(\boldsymbol{r}_1, \boldsymbol{r}_2, \ldots)$$

is well beyond any hope of an exact solution by available techniques. There is a crucially important exception: if V is a *quadratic* function of the variables $\boldsymbol{r}_1, \boldsymbol{r}_2, \ldots$, then a suitable change of these variables will convert the Hamiltonian into a sum of independent Hamiltonians, each of simple harmonic oscillator form. What is more, finding the correct transformation of variables is *exactly* the normal mode problem for the corresponding *classical* system.

We shall not give the details. Instead we shall give, without proof, a useful *recipe* for constructing the solution of the quantum problem once the normal modes for the corresponding classical system have been found. That such a recipe is possible follows from the remarks of the last paragraph, along with the fact that the only really relevant parameter for a simple harmonic oscillator is its frequency (p. 127).

For some applications, it is sufficient to know the energy levels. Having determined the frequencies $\omega_1, \omega_2, \ldots$ of the normal modes of the classical system, we may immediately write the energy levels of the quantum system:

$$\hbar\omega_1(n_1 + \tfrac{1}{2}) + \hbar\omega_2(n_2 + \tfrac{1}{2}) + \ldots$$

where each n_1, n_2, ranges over non-negative integers. For example, the energy eigenvalues of the small internal vibrations of the triatomic molecule of the last section are

$$(n_1 + \tfrac{1}{2})\hbar(K/m)^{\frac{1}{2}} + (n_2 + \tfrac{1}{2})\hbar(3K/m)^{\frac{1}{2}}$$

Each normal mode contributes as if it were an oscillator in its own right, as indeed it is.

Sometimes we need to go further. Then the recipe runs as follows: write down a set of normal modes with amplitudes $A_1, A_2, \ldots$ chosen so that the normal modes have energies respectively equal to $\hbar\omega_1, \hbar\omega_2, \ldots$. Write down the solution which is the superposition of all the normal modes of this set (the phases are actually unimportant, and to fix ideas, we shall use $\eta = 0$ in the example). Finally, replace $\exp(-\mathrm{i}\omega_j t)$ and $\exp(\mathrm{i}\omega_j t)$ by a_j and a_j^+ throughout to obtain the quantum operators which represent the coordinates.

For example, the total energy of the triatomic molecule of the last section is

$$E = \tfrac{1}{2}m\dot{x}^2 + \tfrac{1}{2}m\dot{y}^2 + \tfrac{1}{2}m\dot{z}^2 + \tfrac{1}{2}K(x-y)^2 + \tfrac{1}{2}K(y-z)^2,$$

while the first normal mode, of frequency $\omega_1 = (K/m)^{\frac{1}{2}}$, is (taking $\eta_1 = 0$)

$$x = A_1 \cos \omega_1 t, \qquad y = 0, \qquad z = -A_1 \cos \omega_1 t.$$

For this mode

$$\begin{aligned} E_1 &= (\tfrac{1}{2}m\omega_1^2 + \tfrac{1}{2}m\omega_1^2)A_1^2 \sin^2 \omega_1 t + (\tfrac{1}{2}K + \tfrac{1}{2}K)A_1^2 \cos^2 \omega_1 t, \\ &= m\omega_1^2 A_1^2, \end{aligned}$$

and this will be $\hbar\omega_1$ if we take $A_1 = (\hbar/m\omega_1)^{\frac{1}{2}} = (\hbar^2/mK)^{\frac{1}{4}}$. Similarly, we find that

$$A_2 = (\hbar^2/27mK)^{\frac{1}{4}}.$$

With these values for A_1 and A_2 we write the more general solution (p. 140),

$$\begin{aligned} x &= A_1 \cos \omega_1 t + A_2 \cos \omega_2 t \\ y &= \qquad\qquad\quad -2A_2 \cos \omega_2 t \\ z &= -A_1 \cos \omega_1 t + A_2 \cos \omega_2 t, \end{aligned}$$

and make the substitution of the recipe to obtain the operators

$$\begin{aligned} \mathsf{x} &= A_1 \tfrac{1}{2}(\mathsf{a}_1 + \mathsf{a}_1^+) + A_2 \tfrac{1}{2}(\mathsf{a}_2 + \mathsf{a}_2^+) \\ \mathsf{y} &= \qquad\qquad\qquad\quad -2A_2 \tfrac{1}{2}(\mathsf{a}_2 + \mathsf{a}_2^+) \\ \mathsf{z} &= -A_1 \tfrac{1}{2}(\mathsf{a}_1 + \mathsf{a}_1^+) + A_2 \tfrac{1}{2}(\mathsf{a}_2 + \mathsf{a}_2^+). \end{aligned}$$

To complete the recipe, we need to say something about the algebraic properties of the operators a_j, a_j^+. In fact, extending the relation on p. 130, we have, for each j,

$$[\mathsf{a}_j, \mathsf{a}_j^+] = \mathsf{a}_j\mathsf{a}_j^+ - \mathsf{a}_j^+\mathsf{a}_j = 1.$$

With these exceptions, everything commutes with everything else: that is, for all unequal j and k,

$$\mathsf{a}_j\mathsf{a}_k = \mathsf{a}_k\mathsf{a}_j, \qquad \mathsf{a}_j\mathsf{a}_k^+ = \mathsf{a}_k^+\mathsf{a}_j, \qquad \mathsf{a}_j^+\mathsf{a}_k^+ = \mathsf{a}_k^+\mathsf{a}_j^+.$$

The Hamiltonian is

$$\mathsf{H} = \hbar\omega_1(\mathsf{a}_1^+\mathsf{a}_1+\tfrac{1}{2})+\hbar\omega_2(\mathsf{a}_2^+\mathsf{a}_2+\tfrac{1}{2})+\ldots;$$

algebraic methods may be used as before to show that any eigenvalue of H may be written

$$E_{n_1n_2\ldots} = \hbar\omega_1(n_1+\tfrac{1}{2})+\hbar\omega_2(n_2+\tfrac{1}{2})+\ldots,$$

where $n_1, n_2, \ldots$ are any selection of non-negative integers, and that the corresponding normalized eigenvector is

$$|n_1n_2\ldots\rangle = (\mathsf{a}_1^+)^{n_1}(\mathsf{a}_2^+)^{n_2}\ldots|00\ldots\rangle(n_1!n_2!\ldots)^{-\frac{1}{2}}.$$

The 'lowest' eigenstate $|00\ldots\rangle$ is characterized by

$$\mathsf{a}_j|00\ldots\rangle = 0$$

for every a_j.

PROBLEMS

The three-dimensional isotropic harmonic oscillator has three coordinates, to be regarded as the components of a vector $\boldsymbol{s} = (s_1, s_2, s_3)$. The Hamiltonian is

$$H = \tfrac{1}{2}\hbar\omega(-\nabla_s^2+\boldsymbol{s}^2).$$

1. Show that for solutions of the form $\psi = \psi_1(s_1)\psi_2(s_2)\psi_3(s_3)$ the Schrödinger equation is separable, the eigenfunctions being

$$\psi_{n_1n_2n_3} = \text{const. } H_{n_1}(s_1)H_{n_2}(s_2)H_{n_3}(s_3)\exp(-\tfrac{1}{2}s^2).$$

Show that the energy eigenvalues are $E_n = \hbar\omega(n+\frac{3}{2})$ for any non-negative integer n, and that E_n is $\frac{1}{2}(n+1)(n+2)$-fold degenerate. (This is in fact the analysis into normal modes.)

2. Show that for solutions of the form $\psi = R(s)Y_{lm}(\theta, \phi)$, the Schrödinger equation is separable, the radial equation being

$$ER = \tfrac{1}{2}\hbar\omega\left\{-\frac{1}{s}\frac{\mathrm{d}^2}{\mathrm{d}s^2}(sR)+\frac{l(l+1)}{s^2}R+s^2R\right\}.$$

Show that, if $E = E_n \equiv \hbar\omega(n+\frac{3}{2})$, there is a solution

$$R = f(s)\exp(-\tfrac{1}{2}s^2),$$

where the polynomial $f = \sum_{j=l}^{n} c_js^j$, with adjacent coefficients related by

$$c_j\{j(j+1)-l(l+1)\} = c_{j-2}2(j+2-n).$$

(Since f is an *even* or an *odd* polynomial, the numbers n, l, must be both even or both odd. The degeneracy of E_n is thus

$$(2n+1)+(2n-3)+\ldots+\begin{cases}3 & (n \text{ odd})\\ 1 & (n \text{ even}).\end{cases}$$

(see diagram). Show that this is $\frac{1}{2}(n+1)(n+2)$, as expected from problem 1.)

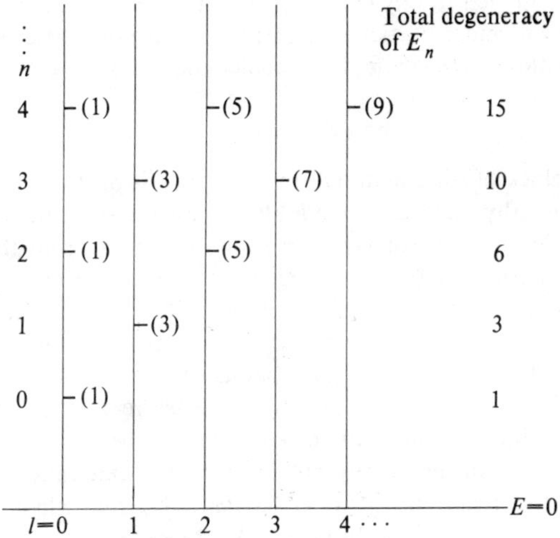

Quanta

We have seen that, in the quantum theory of small oscillations and of normal modes generally, the energy levels have a characteristic structure: apart from the **zero-point energies** $\frac{1}{2}\hbar\omega_1$, $\frac{1}{2}\hbar\omega_2, \ldots$, the typical level comprises a number n_1 of units $\hbar\omega_1$, a number n_2 of units $\hbar\omega_2$, and so on. The fact that we are concerned with integer numbers of basic units (**quanta**, singular **quantum**) which may be counted suggests that these units may be experimentally observed as separate entities: alternatively, that 'objects' seen in the laboratory may be in fact excitations of the normal modes of some underlying oscillatory structure.

The modern view of 'elementary' particles is exactly this. Electrons, nucleons, mesons and the rest are nowadays regarded as quanta in this sense, being excitations of an underlying *quantum field*. The theoretical development of such a field introduces no new quantum principles, and leans very heavily on normal mode techniques; however, field theory is beyond the scope of this book and, apart from a few *ad hoc* results obtained below, will not be dealt with here.

However, it is worth noticing that the phenomenon of radioactive decay needs a development of this kind. An unstable particle, here today, may have gone tomorrow, having decayed into a variety of new particles not previously present. When such a particle is theoretically described as the excitation of a normal mode, its disappearance poses no problems of description: the value of n_j for that mode is simply reduced by one. Out of this there have sprung picturesque names for the operators a_j and a_j^+. The operator a_j^+, applied to any energy eigenstate, increases the value of n_j by one: there is one more particle present; a_j^+ is therefore sometimes called a **creation operator**. Similarly, a_j is an **annihilation** or **destruction operator**.

Phonons

Consider a block of solid material. In so far as it is held together by forces which are basically electrical, such a block may in a sense be regarded as a huge molecule, no different (apart from its hugeness) from the triatomic molecule we have already considered. In particular, we may consider the small interatomic vibrations of the block; these will be more or less harmonic, and may in principle be handled by the normal mode approach. (In practice, if the material is amorphous, it is not easy to analyse the normal modes, as the higher ones tend to be very irregular indeed; a regular crystalline structure is much easier to handle.)

The frequencies of the modes will cover a very wide range. At the low-frequency end there are the everyday acoustic vibrations which, if the block is not too small, can be easily heard. At the other end of the scale are frequencies of much the same size as the vibrational frequencies of small molecules. In spite of this enormous range, there is no essential difference in nature between the normal modes: the excitation of each may be described by the presence of an appropriate number of the corresponding quanta, namely **phonons**.

Though there is no essential difference in *nature* between the extremes, our *attitudes* to them may be very different indeed. At the low-frequency end, the quanta $\hbar\omega$ are so small, and the number present therefore so large that classical mechanics is utterly satisfactory. At the high-frequency end, however, the quanta are large enough for quantum effects to become noticeable. For example, the specific heat of a solid is almost entirely due to the vibrational motion of the constituent atoms. Classical mechanics predicts a value for the specific heat which is good only at high temperatures, but which is somewhat too high (much too high for very hard substances) at room temperatures.

Weber in 1872 was perhaps the first to produce convincing experimental evidence for the anomaly. He discovered that at $-50\,^{\circ}\mathrm{C}$ the specific heat of diamond is only about one-eighth of the classically predicted value. It was in 1906 that Einstein realized that the quantum ideas suggested by Planck for electromagnetic radiation ought equally to apply to atomic vibrations;

in this way, the anomalies have been completely removed. That it was possible to do this twenty years before the development of quantum theory in its final form is because all that was needed for the purpose was the energy level structure, and this structure had already been suggested by Planck for other reasons.

Photons

Maxwell's equations for the electromagnetic field in empty space are linear and homogeneous in the field intensities $\mathscr{E}$ and $\boldsymbol{B}$; therefore their general solution may be expressed as a linear combination of normal modes. The recipe (p. 141) may therefore be applied to give the quantum version of the electromagnetic field, one consequence being the immediate appearance of another type of quantum, the **photon**: each normal mode of angular frequency ω has photons of energy $\hbar\omega$ associated with it. Whether we think of electromagnetic radiation as a collection of oscillators or as a stream of particles depends more than anything else on the particular situation or application we have in mind. In Chapter 1 it was useful to think of the photon as a kind of particle; now we see that we may need to think more flexibly.

It may be worth mentioning that from the modern point of view, phonons and photons have a lot in common, even though the former have an obvious *mechanical* origin which the latter do not. A physicist will even talk of the phonon 'gas' in the solid, as if the phonons were particles with their own mechanics, capable of scattering each other; the underlying structure of the solid is then forgotten for the time being. This contrasts with the attitude of the nineteenth century, when a physicist would not be satisfied until he had some kind of mechanical interpretation for everything. This was especially true of the electromagnetic field, and all kinds of bizarre arrangements of masses, springs, gyroscopes and rubber driving bands were proposed, whose overall behaviour could be described by Maxwell's equations. Nowadays, we are more ready to accept things as they appear to be, having learned not to enquire too closely into what is 'really' happening.

Let us apply the recipe of p. 141 to the normal modes of Maxwell's equations. First, we imagine that the field is confined to a large cube of side L, and adopt *periodic boundary conditions* for which at all times the values of the field intensities on each face of the cube match those on the opposite face. (That such boundary conditions are just as good as more usual ones may be seen by thinking of a cube the size of the solar system: it cannot make any difference, but it is somewhat more convenient.) A plane wave, travelling in the z-direction with wave number k, and plane-polarized in the x-direction has an electric vector

$$\mathscr{E} = A\boldsymbol{\lambda}\cos(kz-\omega t) \quad \text{with} \quad \boldsymbol{\lambda} = (1,0,0) \quad \text{and} \quad \omega = kc.$$

The boundary conditions are satisfied by taking $k = (2\pi/L)n_z$, with n_z any non-negative integer. The recipe requires the amplitude A to be chosen so that this normal mode carries an energy $\hbar\omega$; in fact, the energy density of such a wave is $\varepsilon_0\mathscr{E}^2$, and integrating over the cube gives $\frac{1}{2}\varepsilon_0 A^2 L^3$. Hence

$$A^2 = 2\hbar\omega/\varepsilon_0 L^3.$$

The recipe now immediately gives us the contribution to $\mathscr{E}$ from this normal mode; it is

$$\left(\frac{2\hbar\omega}{\varepsilon_0 L^3}\right)^{\frac{1}{2}} \boldsymbol{\lambda}\tfrac{1}{2}(\mathrm{e}^{ikz}\mathrm{a}+\mathrm{e}^{-ikz}\mathrm{a}^+),$$

where a and a^+ are the annihilation and creation operators for photons of wave-vector $\boldsymbol{k} = (0, 0, k)$, frequency ω, and plane-polarization $\boldsymbol{\lambda}$. (We really ought to write $\mathrm{a}(\boldsymbol{k}, \boldsymbol{\lambda})$, etc, as there is a pair of such operators for *every* normal mode.)

The number of stationary states for this system is truly enormous, and we shall consider only a very small selection which will be useful for our later work, namely those states for which there is at most *one* photon present. First, we have the state of complete darkness, which we represent by the normalized vector $|\text{dark}\rangle$. Then there is the set of states, represented by $|\boldsymbol{k}, \boldsymbol{\lambda}\rangle$, in each of which there is exactly one photon present with wave vector $\boldsymbol{k}$ and polarization $\boldsymbol{\lambda}$. (The vector $\boldsymbol{k}$ needs to be of the form $(2\pi/L)(n_x, n_y, n_z)$ to satisfy the boundary conditions, and $\boldsymbol{\lambda}$ is one or other of two unit vectors, perpendicular to $\boldsymbol{k}$ and to each other.) It is convenient to normalize $|\boldsymbol{k}, \boldsymbol{\lambda}\rangle$ by

$$\langle \boldsymbol{k}, \boldsymbol{\lambda}|\boldsymbol{k}, \boldsymbol{\lambda}\rangle = (L/2\pi)^3$$

since it may then be shown that in the limit of very large L

$$\sum_{k\lambda} |\boldsymbol{k}, \boldsymbol{\lambda}\rangle(2\pi/L)^3\langle \boldsymbol{k}, \boldsymbol{\lambda}| \text{ goes over into } \sum_{\lambda} \int |\boldsymbol{k}, \boldsymbol{\lambda}\rangle\, \mathrm{d}^3\boldsymbol{k}\langle \boldsymbol{k}, \boldsymbol{\lambda}|,$$

as required as a contribution to the usual representation of the unit operator (cf. p. 112).

For the normal mode given explicitly above, $\boldsymbol{k} = (2\pi/L)(0, 0, n_z)$, $\boldsymbol{\lambda} = (1, 0, 0)$; bearing in mind the unconventional normalization of $|\boldsymbol{k}, \boldsymbol{\lambda}\rangle$, we have the matrix elements (problem 3, p. 132)

$$\langle \text{dark}\,|\mathrm{a}|\boldsymbol{k}, \boldsymbol{\lambda}\rangle = (L/2\pi)^{\frac{3}{2}} \quad \text{and} \quad \langle \text{dark}\,|\mathrm{a}^+|\boldsymbol{k}, \boldsymbol{\lambda}\rangle = 0.$$

Hence,

$$\langle \text{dark}\,|\mathscr{E}(\boldsymbol{r})|\boldsymbol{k}, \boldsymbol{\lambda}\rangle = \left(\frac{L}{2\pi}\right)^{\frac{3}{2}}\left(\frac{2\hbar\omega}{\varepsilon_0 L^3}\right)^{\frac{1}{2}} \boldsymbol{\lambda}\tfrac{1}{2}\, \mathrm{e}^{ikz},$$

$$= (\hbar\omega/16\pi^3\varepsilon_0)^{\frac{1}{2}}\boldsymbol{\lambda}\, \mathrm{e}^{ikz}.$$

This is for one normal mode in particular. However, all reference to the original cube has disappeared, so confirming that the boundary conditions are not important. By inspection, we may now generalize this expression to any $\boldsymbol{k}$, obtaining

$$\langle \text{dark}\,|\mathscr{E}(\boldsymbol{r})|\boldsymbol{k},\boldsymbol{\lambda}\rangle = (\hbar\omega/16\pi^3\varepsilon_0)^{\frac{1}{2}}\boldsymbol{\lambda}\exp(\mathrm{i}\boldsymbol{k}\,.\,\boldsymbol{r}),$$

with $\omega = |\boldsymbol{k}|c$, as the matrix element for the electric intensity $\mathscr{E}(\boldsymbol{r})$ between the dark state, and a state with one photon present.

It is not difficult to extend the discussion to include *all* plane waves with a definite polarization. The electric vector of any plane wave with wavevector $\boldsymbol{k}$ and frequency $\omega = c|\boldsymbol{k}|$ may be written

$$\mathscr{E} = \boldsymbol{Z}\exp(\mathrm{i}\boldsymbol{k}\,.\,\boldsymbol{r}-\mathrm{i}\omega t)+\boldsymbol{Z}^*\exp(-\mathrm{i}\boldsymbol{k}\,.\,\boldsymbol{r}+\mathrm{i}\omega t),$$

where the (possibly complex) vector $\boldsymbol{Z}$ fully specifies both the amplitude and the polarization of the wave. Now introduce a vector $\boldsymbol{\lambda} = c\boldsymbol{Z}$ where c is a 'normalizing' factor making $\boldsymbol{\lambda}$ a 'unit' vector:

$$\boldsymbol{\lambda}^*\,.\,\boldsymbol{\lambda} = 1.$$

Then it may be shown that

$$\langle \text{dark}\,|\mathscr{E}(\boldsymbol{r})|\boldsymbol{k},\boldsymbol{\lambda}\rangle = (\hbar\omega/16\pi^3\varepsilon_0)^{\frac{1}{2}}\boldsymbol{\lambda}\exp(\mathrm{i}\boldsymbol{k}\,.\,\boldsymbol{r})$$

exactly as before, except that the vector $\boldsymbol{\lambda}$ (now complex) can describe any kind of definite polarization. We shall need this result for the radiation theory of Chapter 15. Note that $\boldsymbol{\lambda}$ is determined only up to a phase factor, since $|\boldsymbol{k},\boldsymbol{\lambda}\rangle$ on the left is similarly determined.

PROBLEM

The electric vector at any point of a circularly polarized plane wave propagating in the positive z-direction rotates in the *positive* sense about the direction of propagation. Show that the corresponding $\boldsymbol{\lambda}$ is $(1, \mathrm{i}, 0)/2^{\frac{1}{2}}$. (The associated photon has an angular momentum of $\hbar$ in the z-direction. Authors disagree about whether this state should be called left- or right-handed circular polarization; my present count is four-all, with many abstentions! My own predilection is for:

right—angular momentum in the direction of propagation $= +\hbar$

left—angular momentum in direction of propagation $= -\hbar$.

Anyway, the other state of angular momentum $-\hbar$ is associated with the vector $(-1, \mathrm{i}, 0)/2^{\frac{1}{2}}$.)

12. Eigenvalue perturbation theory

The basic idea of a perturbation method

VERY few realistic quantum problems are exactly solvable. In practice, however, a not inconsiderable number of such problems lie 'close' to problems which *are* exactly solvable, and perturbation theory aims to approximate their solution, taking the known solution of the neighbouring problem as a starting point. This rather general remark leads to a host of different applications, and in this book we consider only the approximate evaluation of *eigenvalues* (this chapter and the next) and of *transition probabilities* (chapter 14).

Any perturbation method in quantum mechanics begins with the statement that the Hamiltonian for the problem is the sum of two terms,

$$\mathsf{H} = \mathsf{H}_0 + \mathsf{H}',$$

where the **unperturbed Hamiltonian** H_0 is a simpler operator which we understand adequately, and the **perturbation** H' is in some sense small. The method is then to 'expand' whatever entity interests us as a 'power series' in H'; more exactly, to evaluate the entity to first, second, ... order of smallness of H', as far as we need to go.

A very large part of the basic work may be done in complete generality, and carried out once for all. The resulting *recipes* may then be applied as appropriate to any particular case, usually in a very simple way. We shall use Dirac notation for all the general work, up to the point where the general recipes are obtained. Whether the Dirac notation is retained or not, when these recipes come to be applied, will depend on what is most appropriate to the situation.

The first order shift of a non-degenerate energy level

To keep track of the different orders of smallness we shall introduce a parameter ε:

$$\mathsf{H} = \mathsf{H}_0 + \mathsf{H}'\varepsilon,$$

and later set $\varepsilon = 1$. Suppose that H_0 has a known *isolated non-degenerate* eigenvalue E_0 with a known *normalized* eigenvector $|E_0\rangle$. (The restriction that E_0 be non-degenerate will be removed in chapter 13.) We naturally assume that there is an eigenvalue E and an eigenvector $|\Psi\rangle$ of H in the neighbourhoods of E_0 and $|E_0\rangle$ and, further, that it is possible to expand in powers of ε,

$$E = E_0 + e_1\varepsilon + e_2\varepsilon^2 + \dots$$
$$|\Psi\rangle = |E_0\rangle + |A\rangle\varepsilon + |B\rangle\varepsilon^2 + \dots$$

(Whether these assumptions turn out to be justified will depend on the particular problem under consideration.)

The eigenvalue–eigenvector relation for H,

$$\mathsf{H}|\Psi\rangle = |\Psi\rangle E,$$

may now be written

$$(\mathsf{H}_0 + \mathsf{H}'\varepsilon)(|E_0\rangle + |A\rangle\varepsilon + |B\rangle\varepsilon^2 + \dots)$$
$$= (|E_0\rangle + |A\rangle\varepsilon + |B\rangle\varepsilon^2 + \dots)(E_0 + e_1\varepsilon + e_2\varepsilon^2 + \dots).$$

Equating to zero the coefficients of ε^0, ε^1. $\varepsilon^2, \dots$ and rearranging slightly gives the sequence,

$$\begin{aligned}
&\text{(zeroth order)} \quad (\mathsf{H}_0 - E_0)|E_0\rangle &&= 0\\
&\text{(first order)} \quad (\mathsf{H}_0 - E_0)|A\rangle - |E_0\rangle e_1 &&= -\mathsf{H}'|E_0\rangle\\
&\text{(second order)} \quad (\mathsf{H}_0 - E_0)|B\rangle - |E_0\rangle e_2 &&= -\mathsf{H}'|A\rangle + |A\rangle e_1\\
&&&\vdots
\end{aligned}$$

where each left side contains the first appearance of new items $|A\rangle$, e_1, $|B\rangle$, $e_2 \dots$ while each right side contains items, all of which have occurred earlier. The zeroth equation tells us nothing new: it reasserts that E_0 is an eigenvalue of H_0 with eigenvector $|E_0\rangle$. The succeeding equations allow us to calculate in turn e_1 and $|A\rangle$, e_2 and $|B\rangle, \dots$ as far as we want to go (in principle at least; sometimes there are considerable computational problems).

For the moment, we shall be content with e_1. Premultiply every term in the first order equation by $\langle E_0|$

$$\langle E_0|(\mathsf{H}_0 - E_0)|A\rangle - \langle E_0|E_0\rangle e_1 = -\langle E_0|\mathsf{H}'|E_0\rangle.$$

Now $\langle E_0|(\mathsf{H}_0 - E_0) = 0$ (because $\langle E_0|$ is an eigenvector of H_0) and $\langle E_0|E_0\rangle = 1$ (as $|E_0\rangle$ was chosen to be normalized). Hence the

Recipe: $e_1 = \langle E_0|\mathsf{H}'|E_0\rangle$;

the first order perturbation (or **shift***) of the eigenvalue E_0 is the mean value of H' for the unperturbed state $|E_0\rangle$.*

This result is true for any isolated non-degenerate E_0 in any quantum mechanical situation. It is therefore quite unnecessary to repeat the work of this section in any particular case: we merely apply the recipe.

How good an approximation to E is $E_0 + e_1$? A useful criterion is that it is necessary (though not always sufficient) that e_1 should be small in comparison with the gap between E_0 and the nearest other eigenvalue of H_0. (This, by the way, is why E_0 needs to be an *isolated* eigenvalue.) Why this criterion is necessary will become apparent when we come to consider *second* order perturbation theory.

The almost harmonic oscillator

Consider the problem posed by the Hamiltonian

$$H = -\frac{\hbar^2}{2m}\frac{\mathrm{d}^2}{\mathrm{d}x^2} + V(x), \quad \text{with} \quad V(x) = \tfrac{1}{2}m\omega^2x^2 + \varepsilon_3x^3 + \varepsilon_4x^4,$$

where ε_3 and ε_4 are small parameters. (Such a Hamiltonian may arise, for example, as an improvement on the simple harmonic approximation which we adopted for the almost rigid molecule. The extra terms will then be the continuation of the expansion of $v_l(r_l+\eta)$, p. 136.) It is then natural to take

$$H_0 = -\frac{\hbar^2}{2m}\frac{\mathrm{d}^2}{\mathrm{d}x^2} + \tfrac{1}{2}m\omega^2x^2,$$

$$H' = \varepsilon_3x^3 + \varepsilon_4x^4,$$

and to apply the rule of the last section.

We shall calculate the first order shift of the lowest eigenvalue $\frac{1}{2}\hbar\omega$ of H_0. We need the corresponding normalized eigenfunction (p. 133):

$$\psi_0(x) = (m\omega/\pi\hbar)^{\frac{1}{4}} \exp(-m\omega x^2/2\hbar).$$

The shift is the mean value of H' for this state; that is,

$$\begin{aligned} e_1 &= \int_{-\infty}^{\infty} \mathrm{d}x\psi_0^*(x)H'\psi_0(x) \\ &= (m\omega/\pi\hbar)^{\frac{1}{2}} \int_{-\infty}^{\infty} \mathrm{d}x \exp(-m\omega x^2/\hbar)(\varepsilon_3x^3 + \varepsilon_4x^4) \\ &= 3\varepsilon_4\hbar^2/4m^2\omega^2. \end{aligned}$$

The term ε_3x^3 makes no contribution at first order, as x^3 is an odd function of x. (Values of relevant integrals are given on p. 236.)

The almost harmonic oscillator: algebraic approach

Usually we shall not be content with the energy shift in the ground state, but will need the shift for higher eigenvalues also. This will involve the calculation of the mean value of H' for wavefunctions having the Hermite polynomials (p. 133) as factors. The Hermite polynomials have an extensive theory which allows this to be done; it is nevertheless quite troublesome if H' involves higher powers of x.

By comparison, the algebraic approach is very direct, as we may expect. The classical solution $x = A\cos\omega t$ has a total energy

$$\tfrac{1}{2}m\dot{x}^2 + \tfrac{1}{2}m\omega^2 x^2 = \tfrac{1}{2}m\omega^2 A^2,$$

and this will be $\hbar\omega$ provided $A^2 = 2\hbar/m\omega$. The recipe on p. 141 now leads us to write

$$\mathsf{x} = \sqrt{(2\hbar/m\omega)}\,.\tfrac{1}{2}(\mathsf{a}+\mathsf{a}^+)$$

and hence, for example,

$$\varepsilon_4\mathsf{x}^4 = \frac{\hbar^2\varepsilon_4}{4m^2\omega^2}(\mathsf{a}+\mathsf{a}^+)^4.$$

The contribution of this term to the shift in the nth eigenvalue is then $\langle n|\varepsilon_4\mathsf{x}^4|n\rangle$.

The expression on the right has 16 terms, each with four factors a or a^+ in a variety of different orders. (Since a and a^+ do not commute, order is important.) Only terms with two as and two a^+s yield non-zero contributions; they are

$$\langle n|\varepsilon_4\mathsf{x}^4|n\rangle = \frac{\hbar^2\varepsilon_4}{4m^2\omega^2}\langle n|(\mathsf{aaa^+a^+} + \mathsf{aa^+aa^+} + \mathsf{a^+aaa^+} + \mathsf{aa^+a^+a} + \mathsf{a^+aa^+a}$$
$$+\mathsf{a^+a^+aa})|n\rangle.$$

Each term may be evaluated on the following lines (see p. 131):

$$\begin{aligned}\langle n|\mathsf{aaa^+a^+}|n\rangle &= \langle n|\mathsf{aaa^+}|n+1\rangle(n+1)^{\frac{1}{2}}\\ &= \langle n|\mathsf{aa}|n+2\rangle(n+2)^{\frac{1}{2}}(n+1)^{\frac{1}{2}}\\ &= \langle n|\mathsf{a}|n+1\rangle(n+2)^{\frac{1}{2}}(n+2)^{\frac{1}{2}}(n+1)^{\frac{1}{2}}\\ &= \langle n|n\rangle(n+1)^{\frac{1}{2}}(n+2)^{\frac{1}{2}}(n+2)^{\frac{1}{2}}(n+1)^{\frac{1}{2}}\\ &= (n+1)(n+2).\end{aligned}$$

Applying this technique to the other five terms also, we get

$$\begin{aligned}\langle n|\varepsilon_4\mathsf{x}^4|n\rangle &= \frac{\hbar^2\varepsilon_4}{4m^2\omega^2}\{(n+1)(n+2)+(n+1)^2\\ &\qquad +n(n+1)+n(n+1)+n^2+n(n-1)\}\\ &= \frac{\hbar^2\varepsilon_4}{4m^2\omega^2}(6n^2+6n+3);\end{aligned}$$

this gives the first order shift for *any* eigenvalue $\hbar\omega(n+\frac{1}{2})$. Setting $n = 0$ gives the result of the last section.

The necessary criterion that any shift should be reasonably accurate is that it should be small compared to the gap $\hbar\omega$ between levels, i.e. that

$$\frac{\hbar \varepsilon_4}{4m^2\omega^3}(6n^2+6n+3) \ll 1.$$

This criterion may always be violated by taking n large enough; thus the approximation will be good (if at all) for the lower levels.

PROBLEMS

1. Show that the potential v_l for the almost rigid diatomic molecule (p. 136) is

$$v_l(r_l+\eta) = g\left\{\frac{-1}{2r_l}+\frac{\eta^2}{2r_l^3}+\frac{\eta^3}{r_l^4}+\frac{3\eta^4}{2r_l^5}+0(\eta^5)\right\}.$$

Hence show that the energy levels are approximately

$$E_{n'l} = -\frac{g}{2r_l}+\hbar\left(\frac{g}{\mu r_l^3}\right)^{\frac{1}{2}}(n'+\tfrac{1}{2})+\frac{3\hbar^2}{8\mu r_l^2}(6n'^2+6n'+3)$$

2. Using the algebraic approach, verify for all n that

$$\langle n|\mathsf{x}^3|n\rangle = 0.$$

The second order shift of an isolated eigenvalue

In most applications of eigenvalue perturbation theory, if the first order correction e_1 is non-zero, it is usually unnecessary to proceed to the calculation of e_2. However, there are many situations in physics where for some reason (usually symmetry) the first order correction e_1 turns out to be exactly zero, and the energy shift is dominated by the value of e_2. Let us now consider how it may be evaluated. Here again it is possible to derive a general recipe once and for all.

The first equation in the sequence (p. 149) to contain e_2 is

$$(\mathsf{H}_0-E_0)|\mathsf{B}\rangle-|E_0\rangle e_2 = -\mathsf{H}'|\mathsf{A}\rangle+|\mathsf{A}\rangle e_1.$$

Premultiply each term by $\langle E_0|$ to obtain the

Recipe: $e_2 = \langle E_0|\mathsf{H}'|\mathsf{A}\rangle-\langle E_0|\mathsf{A}\rangle e_1$

(compare with the derivation of e_1, p. 149). This gives a general expression for e_2, provided that e_1 and $|\mathsf{A}\rangle$ have been first obtained.

Now e_1 is given on p. 149, and it remains to find $|\mathsf{A}\rangle$. In fact, $|\mathsf{A}\rangle$ is a solution of the *inhomogeneous* linear equation

$$(\mathsf{H}_0-E_0)|\mathsf{A}\rangle = -(\mathsf{H}'-e_1)|E_0\rangle.$$

The solution is not unique, since another solution may be obtained by adding on any constant multiple of $|E_0\rangle$. It is usual to make the solution unique by requiring in addition that

$$\langle E_0|\mathsf{A}\rangle = 0;$$

if this is done, the expression for e_2 simplifies to

$$e_2 = \langle E_0|\mathsf{H}'|\mathsf{A}\rangle.$$

It remains to consider how to solve the equation for $|\mathsf{A}\rangle$. This equation differs from the original Schrödinger equation for H_0 (which we are presumed to understand well) only by the new feature of the non-zero **source** term on the right side. We may therefore expect that its solution will proceed along similar lines: we shall call this the *direct approach*, and give some straightforward examples later.

Sometimes the direct approach is not convenient, and it is then usual to use an *expansion method*, in which $|\mathsf{A}\rangle$ is expressed as a superposition of normalized eigenvectors of H_0 (effectively the H_0-representation):

$$|\mathsf{A}\rangle = \sum_{n\neq 0} |E_n\rangle\langle E_n|\mathsf{A}\rangle.$$

The term in $|E_0\rangle$ is excluded to ensure that $\langle E_0|\mathsf{A}\rangle = 0$. (By the way, even though E_0 is to be isolated and non-degenerate, the other eigenvalues E_n need not be, and the expansion of $|\mathsf{A}\rangle$ may involve integration in addition to summation; but this need not detain us.) The equation to be satisfied by $|\mathsf{A}\rangle$ is now premultiplied by $\langle E_n|$ for each $n \neq 0$, giving

$$\langle E_n|(\mathsf{H}_0 - E_0)|\mathsf{A}\rangle = -\langle E_n|\mathsf{H}'|E_0\rangle$$

(note that the term with e_1 drops out, by orthogonality). Hence

$$(E_n - E_0)\langle E_n|\mathsf{A}\rangle = -\langle E_n|\mathsf{H}'|E_0\rangle \qquad (n \neq 0)$$

or

$$\langle E_n|\mathsf{A}\rangle = -\frac{\langle E_n|\mathsf{H}'|E_0\rangle}{E_n - E_0}. \qquad (n \neq 0)$$

These are the coefficients in the expansion of $|\mathsf{A}\rangle$, which is therefore

$$|\mathsf{A}\rangle = -\sum_{n\neq 0} |E_n\rangle \frac{\langle E_n|\mathsf{H}'|E_0\rangle}{E_n - E_0}.$$

This result is inserted in the expression for e_2 to give

$$e_2 = \langle E_0|\mathsf{H}'|\mathsf{A}\rangle = -\sum_{n\neq 0} \langle E_0|\mathsf{H}'|E_n\rangle \frac{\langle E_n|\mathsf{H}'|E_0\rangle}{E_n - E_0}$$

or an alternative explicit

$$\textbf{Recipe:} \quad e_2 = -\sum_{n\neq 0} \frac{|\langle E_0|\mathsf{H}'|E_n\rangle|^2}{E_n - E_0}.$$

An occasionally useful check on the algebra in any application is the fact that, when E_0 is the *lowest* eigenvalue of H_0, then e_2 is necessarily negative.

The reason why the zeroth approximation E_0 must be an isolated nondegenerate eigenvalue of H_0 is now apparent. If this were not so, the sum over n would include denominators $E_n - E_0$ whose values will be very small, and the perturbation recipes would no longer make sense. In the next chapter we shall introduce the important modifications that are needed when E_0 is degenerate. On the other hand, it is never possible to apply these techniques when E_0 belongs to a *continuous* range of eigenvalues.

It is a remarkable fact that the *third* order correction e_3 does *not* require a knowledge of $|B\rangle$ for its evaluation; the formula depends only on $|A\rangle$, and is

$$e_3 = \langle A|H'|A\rangle - \langle A|A\rangle e_1$$

This result is not often used, however, and we do not prove it here.

The almost harmonic oscillator: second order corrections

Consider the system with Hamiltonian

$$\mathsf{H} = \mathsf{H}_0 + \mathsf{H}' \leftarrow \left\{ -\frac{\hbar^2}{2m}\frac{d^2}{dx^2} + \tfrac{1}{2}m\omega^2 x^2 \right\} + \varepsilon_3 x^3.$$

We have already seen (p. 150) that the perturbation $\mathsf{H}' = \varepsilon_3 \mathsf{x}^3$ has no effect in first order in ε_3; here we calculate its effect in second order, using first the algebraic approach.

First we note that

$$\mathsf{H}' = \varepsilon_3(\hbar/2m\omega)^{\frac{3}{2}}(\mathsf{a}+\mathsf{a}^+)^3$$

(see p. 151), and calculate $\langle n_0|\mathsf{H}'|n\rangle$ for all $n \neq n_0$, with the explicit recipe for e_2 in mind (p. 153). Now it happens that H' expands into a sum of 8 terms, each having three factors a or a^+ in some order, and it is easy to verify that $\langle n_0|\mathsf{H}'|n\rangle$ is *zero*, unless $n = n_0 \pm 1$ or $n_0 \pm 3$. The series for e_2 therefore collapses to four terms, and is straightforward to evaluate exactly.

Let us evaluate e_2 for the ground state ($n_0 = 0$). Matters are now even simpler, as only two terms survive. We have, as the only possible non-zero contributions,

$$\langle 0|\mathsf{H}'|3\rangle = \varepsilon_3(\hbar/2m\omega)^{\frac{3}{2}}\langle 0|\mathsf{aaa}|3\rangle = \varepsilon_3(\hbar/2m\omega)^{\frac{3}{2}}\sqrt{6}$$

and

$$\begin{aligned}\langle 0|\mathrm{H}'|1\rangle &= \varepsilon_3(\hbar/2m\omega)^{\frac{3}{2}}\langle 0|(\mathrm{aaa}^+ + \mathrm{aa}^+\mathrm{a} + \mathrm{a}^+\mathrm{aa})|1\rangle \\ &= \varepsilon_3(\hbar/2m\omega)^{\frac{3}{2}}(2+1+0) \\ &= 3\varepsilon_3(\hbar/2m\omega)^{\frac{3}{2}}.\end{aligned}$$

These lead immediately to

$$\begin{aligned}e_2 &= -\frac{|\langle 0|\mathrm{H}'|1\rangle|^2}{E_1 - E_0} - \frac{|\langle 0|\mathrm{H}'|3\rangle|^2}{E_3 - E_0} \\ &= -\varepsilon_3^2(\hbar/2m\omega)^3\left(\frac{9}{\hbar\omega} + \frac{6}{3\hbar\omega}\right) \quad (\text{using } E_n = \hbar\omega(n+\tfrac{1}{2})) \\ &= -11\varepsilon_3^2\hbar^2/8m^3\omega^4,\end{aligned}$$

which is the required second-order shift.

In the more general case of

$$\mathrm{H}' = \varepsilon_3 \mathrm{x}^3 + \varepsilon_4 \mathrm{x}^4,$$

we may need to ask which is more important: the first-order shift due to ε_4 or the second-order shift due to ε_3. A measure of their relative importance is the quotient

$$\left(\frac{11}{8}\frac{\varepsilon_3^2\hbar^2}{m^3\omega^4}\right)\Bigg/\left(\frac{4}{3}\frac{\hbar^2\varepsilon_4}{m^2\omega^2}\right) = \frac{11}{6m\omega^2}\cdot\frac{\varepsilon_3^2}{\varepsilon_4}.$$

PROBLEMS

1. Show that

$$\begin{aligned}\langle n|(\mathrm{a}+\mathrm{a}^+)^3|n+3\rangle &= \{(n+1)(n+2)(n+3)\}^{\frac{1}{2}}, \\ \langle n|(\mathrm{a}+\mathrm{a}^+)^3|n+1\rangle &= 3(n+1)^{\frac{3}{2}} \\ \langle n|(\mathrm{a}+\mathrm{a}^+)^3|n-1\rangle &= 3n^{\frac{3}{2}} \\ \langle n|(\mathrm{a}+\mathrm{a}^+)^3|n-3\rangle &= \{n(n-1)(n-2)\}^{\frac{1}{2}}\end{aligned}$$

Hence, for state n, evaluate

$$e_2 = -(\varepsilon_3^2\hbar^2/8m^3\omega^4)(30n^2+30n+11).$$

2. Show that in problem 1, p. 152, the second order contribution of the term in η^3 to $E_{n'l}$ is

$$-\frac{\hbar^2}{8\mu r_l^2}(30n'^2+30n'+11),$$

comparable in size to the first order effect of the term in η^4.

The almost harmonic oscillator: the direct approach

We shall repeat the evaluation of e_2 of the last section, this time using the direct approach (p. 153) for the purpose of comparison. We use the s-representation throughout. First, we summarize the procedure.

The Hamiltonian in the s-representation is (p. 154)

$$\mathsf{H} = \mathsf{H}_0 + \mathsf{H}' \leftarrow \hbar\omega\left(-\frac{1}{2}\frac{d^2}{ds^2} + \frac{1}{2}s^2\right) + \varepsilon_3 x_0^3 s^2.$$

We take as $|E_0\rangle$ the ground state of H_0, represented by (p. 132)

$$\langle s|E_0\rangle = \pi^{-\frac{1}{4}} \exp\left(-\tfrac{1}{2}s^2\right), \quad \text{with eigenvalue } E_0 = \tfrac{1}{2}\hbar\omega.$$

We require to find $|\mathrm{A}\rangle$, satisfying

$$(\mathsf{H}_0 - E_0)|\mathrm{A}\rangle = (-\mathsf{H}' + e_1)|E_0\rangle.$$

Premultiplying by $\langle s|$ throughout yields an equation for the representative $\langle s|\mathrm{A}\rangle$:

$$\hbar\omega\left(-\frac{1}{2}\frac{d^2}{ds^2} + \frac{1}{2}s^2 - \frac{1}{2}\right)\langle s|\mathrm{A}\rangle = -\varepsilon_3 x_0^3 s^3 \langle s|E_0\rangle$$

(remembering that in this case $e_1 = 0$). We must also satisfy

$$0 = \langle E_0|\mathrm{A}\rangle = \int_{-\infty}^{\infty} \langle E_0|s\rangle \, ds \langle s|\mathrm{A}\rangle.$$

Having found $\langle s|\mathrm{A}\rangle$, it remains to evaluate

$$e_2 = \langle E_0|\mathsf{H}'|\mathrm{A}\rangle = \int_{-\infty}^{\infty} ds \langle E_0|s\rangle \,.\, \varepsilon_3 x_0^3 s^3 \,.\, \langle s|\mathrm{A}\rangle.$$

Now for the details. We first need to solve the equation

$$\hbar\omega\left(-\frac{1}{2}\frac{d^2}{ds^2} + \frac{1}{2}s^2 - \frac{1}{2}\right)\langle s|\mathrm{A}\rangle = -\varepsilon_3 x_0^3 \pi^{-\frac{1}{4}} \,.\, s^3 \exp\left(-\tfrac{1}{2}s^2\right).$$

Making the familiar substitution (p. 128)

$$\langle s|\mathrm{A}\rangle = f_1(s) \exp\left(-\tfrac{1}{2}s^2\right)$$

leads to an equation for f_1,

$$\hbar\omega\left(-\tfrac{1}{2}f_1'' + s f_1'\right) = -\varepsilon_3 x_0^3 \pi^{-\frac{1}{4}} s^3.$$

In the light of previous experience with the Schrödinger equation for the simple harmonic oscillator, it is no surprise to discover that $f_1(s)$, on account of the boundary requirements, must be a polynomial in s,

$$f_1(s) = -\left(\varepsilon_3 x_0^3/\pi^{\frac{1}{4}}\hbar\omega\right)\left(s + \tfrac{1}{3}s^3\right).$$

It follows that the first-order correction $|A\rangle$ to the eigenvector $|E_0\rangle$ is represented by

$$\langle s|A\rangle = -(\varepsilon_3 x_0^3/\pi^{\frac{1}{4}}\hbar\omega)(s+\tfrac{1}{3}s^3)\exp(-\tfrac{1}{2}s^2).$$

(The requirement $\langle E_0|A\rangle = 0$ is already satisfied, because the integrand in $\int\langle E_0|s\rangle\, ds\langle s|A\rangle$ is an odd function of s.)

We may now evaluate e_2;

$$e_2 = -\int_{-\infty}^{\infty} ds\,.\,\pi^{-\frac{1}{4}}\exp(-\tfrac{1}{2}s^2)\,.\,\varepsilon_3 x_0^3 s^3\,.$$

$$(\varepsilon_3 x_0^3/\pi^{\frac{1}{4}}\hbar\omega)(s+\tfrac{1}{3}s^3)\exp(-\tfrac{1}{2}s^2)$$

$$= -11\varepsilon_3^2\hbar^2/8m^3\omega^4,$$

on evaluating the standard integrals (p. 236). This agrees completely with the result of the previous section.

The apparent longwindedness of the 'direct' approach is a result of the special nature of the simple harmonic oscillator, for which the explicit series for e_2 is rather short. Usually, the series is infinite, and troublesome to sum, and the direct method may then offer distinct advantages.

PROBLEMS

1. Verify that $f_1(s)$ satisfies the inhomogeneous differential equation.
2. Show, by the method of this section, that the second order shift in the energy level $E_1 = \frac{3}{2}\hbar\omega$ is $-71\varepsilon_3^2\hbar^2/8m^3\omega^4$.

The polarizability of the ground state of the hydrogen atom

The mean electric dipole moment of the ground state of the hydrogen atom is zero. This does not imply that the atom is not affected by an external electric field; far from it. Because the atom contains two charged particles which are separately affected by the external field, we may expect it to be *deformed* in some way by the field, acquiring in the process an *induced* dipole moment $\alpha\mathscr{E}$, proportional (for small fields at least) to the field $\mathscr{E}$. The constant α is the **polarizability** of the atom. The simplest way to calculate α is to calculate the second-order energy shift of the ground state in the presence of a field $\mathscr{E}$; this is $-\frac{1}{2}\alpha\mathscr{E}^2$.

As in the last section, we begin with a summary of the procedure. The Hamiltonian for this problem is (in the $\boldsymbol{r}$-representation)

$$\mathsf{H} = \mathsf{H}_0 + \mathsf{H}' \leftarrow \left\{-\frac{\hbar^2}{2m}\nabla^2 - \frac{e^2}{4\pi\varepsilon_0 r}\right\} + e\boldsymbol{\mathscr{E}}\,.\,\boldsymbol{r}.$$

(We assume the proton to be *fixed* at O; thus $\eta = 1$.) We take as $|E_0\rangle$, the state whose polarizability is sought, the ground state represented by

$$\langle \boldsymbol{r}|E_0\rangle = \pi^{-\frac{1}{2}}a^{-\frac{3}{2}}\exp(-r/a_0),$$

with eigenvalue $E_0 = -\hbar^2/2ma_0^2$ (p. 100). The equation to be satisfied by $|\mathrm{A}\rangle$ (or, rather, $\langle \boldsymbol{r}|\mathrm{A}\rangle$ in the $\boldsymbol{r}$-representation) is

$$\langle \boldsymbol{r}|(\mathsf{H}_0 - E_0)|\mathrm{A}\rangle = \langle \boldsymbol{r}|(-\mathsf{H}' + e_1)|E_0\rangle,$$

that is,

$$\left(-\frac{\hbar^2}{2m}\nabla^2 - \frac{e^2}{4\pi\varepsilon_0 r} + \frac{\hbar^2}{2ma_0^2}\right)\langle \boldsymbol{r}|\mathrm{A}\rangle = -e\mathscr{E}.\boldsymbol{r}\langle \boldsymbol{r}|E_0\rangle,$$

(since $e_1 = 0$ in this case). The extra condition

$$0 = \langle E_0|\mathrm{A}\rangle = \int\limits_{\text{all space}} \mathrm{d}^3\boldsymbol{r}\langle E_0|\boldsymbol{r}\rangle\langle \boldsymbol{r}|\mathrm{A}\rangle$$

must also be satisfied. The desired energy shift is then given by

$$e_2 = \langle E_0|\mathsf{H}'|\mathrm{A}\rangle = \int\limits_{\text{all space}} \mathrm{d}^3\boldsymbol{r}\langle E_0|\boldsymbol{r}\rangle e\mathscr{E}.\boldsymbol{r}\langle \boldsymbol{r}|\mathrm{A}\rangle.$$

The foregoing summary is no more complicated than that of the last section. The fact that the details which now follow are more ramified follows simply from the more realistic nature of the application. The equation for $\langle \boldsymbol{r}|\mathrm{A}\rangle$ will be solved in spherical polar coordinates with axis along $\mathscr{E}$; thus $\mathscr{E}.\boldsymbol{r} = \mathscr{E}r\cos\theta$. The right side of the equation now includes a factor $\cos\theta$; we therefore expect that $\langle \boldsymbol{r}|\mathrm{A}\rangle$ must include such a factor, and we try for a solution of the form

$$\langle \boldsymbol{r}|\mathrm{A}\rangle = F(r)\cos\theta.$$

Now the work on the separation of variables (p. 80) shows that

$$\nabla^2(F(r)\cos\theta) = \left(\frac{1}{r^2}\frac{\partial}{\partial r}\left(r^2\frac{\partial F}{\partial r}\right) - \frac{l(l+1)F}{r^2}\right)\cos\theta$$

with $l = 1$. Hence the equation to be satisfied by $F(r)$ is

$$-\frac{\hbar^2}{2m}\frac{1}{r^2}\frac{\mathrm{d}}{\mathrm{d}r}\left(r^2\frac{\mathrm{d}F}{\mathrm{d}r}\right) + \frac{\hbar^2 F}{mr^2} - \frac{e^2 F}{4\pi\varepsilon_0 r} + \frac{\hbar^2 F}{2ma_0^2} = -e\mathscr{E}\pi^{-\frac{1}{2}}a_0^{-\frac{3}{2}}re^{-r/a_0}.$$

The problem has now become one-dimensional. Previous experience with the radial Schrödinger equation for the hydrogen atom (p. 96) suggests that, on account of the boundary requirements,

$$F(r) = \exp(-r/a_0) \times (\text{a polynomial in } r),$$

and it is straightforward to show that the equation is satisfied by

$$F(r) = -(e\mathscr{E}ma_0^{\frac{1}{2}}/\hbar^2\pi^{\frac{1}{2}})(r+r^2/2a_0)\exp(-r/a_0).$$

The rest is a matter of evaluating an integral for e_2:

$$\begin{aligned} e_2 &= \langle E_0|\mathsf{H}'|\mathrm{A}\rangle \\ &= \int r^2 \sin\theta \,\mathrm{d}r\,\mathrm{d}\theta\,\mathrm{d}\phi\,.\,\pi^{-\frac{1}{2}}a_0^{-\frac{3}{2}}\exp(-r/a_0)\,.\,e\mathscr{E}r\cos\theta\,.\,F(r)\cos\theta \\ &= -(e^2\mathscr{E}^2m/\pi\hbar^2 a_0)\int_0^\infty \mathrm{d}r\, r^3\left(r+\frac{r^2}{2a_0}\right)\exp(-2r/a_0) \\ &\quad\times\int_0^\pi \sin\theta\,\mathrm{d}\theta\cos^2\theta\,.\int_0^{2\pi}\mathrm{d}\phi \\ &= -\tfrac{9}{4}(4\pi\varepsilon_0)\mathscr{E}^2a_0^3, \end{aligned}$$

on evaluating the integrals and using $a_0 = 4\pi\varepsilon_0\hbar^2/me^2$. The polarizability is therefore

$$\alpha = \tfrac{9}{2}(4\pi\varepsilon_0)a_0^3.$$

In what way is the atom deformed by the field? This is answered by the first-order change in $|E_0\rangle$, namely $|\mathrm{A}\rangle$. Now

$$\langle \mathbf{r}|\mathrm{A}\rangle = F(r)\cos\theta = -\langle \mathbf{r}|E_0\rangle\,.\,(4\pi\varepsilon_0\mathscr{E}a_0/e)(r+r^2/2a_0)\cos\theta,$$

and the ground state $|E\rangle$ of the *full* Hamiltonian H is given to first order by

$$\psi(\mathbf{r}) = \langle \mathbf{r}|E\rangle = \langle \mathbf{r}|E_0\rangle\{1-(4\pi\varepsilon_0\mathscr{E}a_0/e)(r+r^2/2a_0)\cos\theta+0(\mathscr{E}^2)\}.$$

This expression will be a useful guide in a later chapter (p. 211). A pictorial representation of $|\psi|^2$ is to be found in diagram (d), p. 233.

PROBLEMS

1. Show that the second order shift of the ground state of a hydrogen atom in the field of a proton at a considerable distance $R(\gg a_0)$ is

$$-\frac{9\hbar^2}{4m_e}\frac{a_0^4}{R^4}.$$

Deduce that there is an inverse-fifth attractive force between the proton and the atom. What would the force be if the proton were replaced by an antiproton, with charge $-e$?

2. Show that with the scaling of the problem on p. 100, the first order perturbed ground state wavefunction is

$$\psi(\boldsymbol{\rho}) = \pi^{-\frac{1}{2}}\{e^{-\rho}+c(\rho+\tfrac{1}{2}\rho^2)e^{-\rho}\cos\theta\},$$

where c is a certain constant.

13. Eigenvalue perturbation theory: the degenerate case

What if E_0 is degenerate?

LET us consider again the equations for the zeroth and first order perturbation (p. 149),

$$(\mathsf{H}_0 - E_0)|E_0\rangle = 0,$$

$$(\mathsf{H}_0 - E_0)|\mathsf{A}\rangle - |E_0\rangle e_1 = -\mathsf{H}'|E_0\rangle.$$

Because we earlier required E_0 to be non-degenerate, the state $|E_0\rangle$ was then essentially unique. If E_0 is a degenerate level, the problem is: which of the many possibilities for $|E_0\rangle$ are we to choose as the first term in the perturbation expansion? We shall find that the choice is not arbitrary.

Suppose that the isolated eigenvalue E_0 is m-fold degenerate; that is, we can find a set of m normalized mutually orthogonal eigenvectors of H_0 (which we label with j or k)

$$|E_0 j\rangle, \quad j = 1, 2, \ldots, m, \quad \text{with} \quad \langle E_0 j|E_0 k\rangle = \delta_{jk}.$$

Then the eigenvector $|E_0\rangle$ which we want must be a superposition of the vectors of this set,

$$|E_0\rangle = \sum_{j=1}^{m} |E_0 j\rangle c_j,$$

where, in order that $|E_0\rangle$ should be normalized, the coefficients $c_1, \ldots c_m$ must satisfy

$$\sum_{j=1}^{m} |c_j|^2 = 1.$$

Premultiply each term in the first order equation by each $\langle E_0 j|$ in turn, and note that $\langle E_0 j|(\mathsf{H} - E_0) = 0$; we obtain the m equations

$$\langle E_0 j|E_0\rangle e_1 = \langle E_0 j|\mathsf{H}'|E_0\rangle \quad \text{for} \quad j = 1, 2, \ldots m.$$

Replacing $|E_0\rangle$ by $\sum |E_0 k\rangle c_k$ now gives

$$c_j e_1 = \sum_{k=1}^{m} \langle E_0 j|\mathsf{H}'|E_0 k\rangle c_k$$

or, writing $V_{jk} = \langle E_0 j|\mathsf{H}'|E_0 k\rangle$,

$$\sum_{k=1}^{m} V_{jk} c_k = e_1 c_j.$$

These equations, for $j = 1, 2, \ldots m$, state that e_1 *is an eigenvalue of the matrix* V whose elements are V_{jk}, while the coefficients $c_1 \ldots c_m$ are the components of the corresponding eigenvector. Hence, when the unperturbed eigenvalue E_0 is m-fold degenerate, we have the

Recipe: (i) For the eigenvalue E_0, find an orthonormal set $|E_0 1\rangle, \ldots, |E_0 m\rangle$ of eigenvectors of H_0;
(ii) Evaluate the m^2 **matrix elements** $V_{jk} = \langle E_0 j|\mathsf{H}'|E_0 k\rangle$;
(iii) The eigenvalues e_1 of the **secular matrix** V give the possible first-order shifts of the eigenvalue E_0.

Since V may have as many as m distinct eigenvalues, the degenerate eigenvalue E_0 may be expected to separate into m distinct components: the degeneracy may be **lifted** (see p. 104). If not all of the eigenvalues of V are distinct, then the splitting is not complete; and the degeneracy is **partially lifted**.

By the way, when $m = 1$, the matrix V collapses to a single element, and has just one eigenvalue

$$e_1 = \langle E_0 1|\mathsf{H}'|E_0 1\rangle.$$

This is, of course, exactly the earlier recipe of p. 149.

The Stark effect for the hydrogen atom

The first excited level ($n = 2$) of the hydrogen atom has a *fourfold* degeneracy (p. 104) which is observed to be partially lifted by a uniform constant electric field. This is the **Stark effect**; we may investigate it with the help of the recipe we have just given.

The first step in the recipe is to select four orthonormal statevectors belonging to the level $n = 2$; these are conveniently specified by their quantum numbers n, l, and m (p. 99), and are

$$|nlm\rangle: \; |200\rangle \quad |211\rangle \quad |210\rangle \quad |21-1\rangle.$$

Their wavefunctions in the $\mathbf{r}$-representation are

$$\langle r\theta\phi|nlm\rangle = \psi_{nlm}(r\theta\phi),$$

and these may be assembled from the information on pp. 99, 81. (Of course, $Z = 1$ in this case.)

Next, we need to evaluate the sixteen elements of the secular matrix V. This is always the most tedious stage in calculations of this kind, since it is usual to have to evaluate a lengthy sequence of quite complex integrals. However, it is useful to remember that degeneracy almost always arises on account of some *symmetry* of H_0, and that this same symmetry may well guarantee that a large number of the integrals are zero; it is usually obvious which these are. Thus the calculation may not be as formidable as it seems.

The perturbation due to the uniform field $\mathscr{E} = (0, 0, \mathscr{E})$ is the potential energy of the electric dipole moment $-e\mathbf{r}$ of the atom in the field; this is

$$\mathsf{H}' = e\boldsymbol{\mathscr{E}} \cdot \mathbf{r} = e\mathscr{E}z = e\mathscr{E}r\cos\theta.$$

A typical matrix element becomes, when evaluated in the $r\theta\phi$-representation,

$$\begin{aligned}\langle nlm|\mathsf{z}|n'l'm'\rangle &= \iiint r^2\,\mathrm{d}r\,\sin\theta\,\mathrm{d}\theta\,\mathrm{d}\phi\,\psi^*_{nlm}(r\theta\phi)r\cos\theta\,\psi_{n'l'm'}(r\theta\phi)\\ &= \int_0^\infty r^3\,\mathrm{d}r R_{nl}(r)R_{n'l'}(r)\,.\iint \sin\theta\cos\theta\,\mathrm{d}\theta\,\mathrm{d}\phi\,Y^*_{lm}(\theta\phi)Y_{l'm'}(\theta\phi)\\ &= J(nl, n'l')\,.\,K_z(lm, l'm').\end{aligned}$$

It is useful to list systematically the values of the integrals we may have to use. For example, for

$$J(nl, n'l') = \int_0^\infty r^3\,\mathrm{d}r R_{nl}(r)R_{n'l'}(r),$$

substituting the expressions for R_{nl} (p. 99) gives straightforwardly enough (with Z and η both set to 1; see also p. 236)

$$\begin{aligned}J(20, 20) &= 6a_0,\\ J(20, 21) = J(21, 20) &= -3\sqrt{3}a_0,\\ J(21, 21) &= 5a_0.\end{aligned}$$

Also, for

$$K_z(lm, l'm') = \int_0^\pi \sin\theta\cos\theta\,\mathrm{d}\theta\int_0^{2\pi}\mathrm{d}\phi\,Y^*_{lm}(\theta, \phi)Y_{l'm'}(\theta, \phi),$$

of the sixteen combinations of $lml'm'$ which are of interest, fourteen are zero on account of obvious antisymmetries in their integrands; this is a direct consequence of the spherical symmetry of H_0. The two survivors are

$$K_z(10, 00) = K_z(00, 10) = 1/\sqrt{3}.$$

It follows that the secular matrix has just two non-zero components,

$$\begin{aligned}\langle 200|\mathsf{H}'|210\rangle &= e\mathscr{E}J(20, 21)K_z(00, 10) = -3e\mathscr{E}a_0,\\ \langle 210|\mathsf{H}'|200\rangle &= -3e\mathscr{E}a_0.\end{aligned}$$

The matrix itself is therefore

$$V = \begin{array}{c} (200) \\ (211) \\ (210) \\ (21-1) \end{array} \overset{\begin{array}{cccc} (200) & (211) & (210) & (21-1) \end{array}}{\begin{pmatrix} 0 & 0 & -3e\mathscr{E}a_0 & 0 \\ 0 & 0 & 0 & 0 \\ -3e\mathscr{E}a_0 & 0 & 0 & 0 \\ 0 & 0 & 0 & 0 \end{pmatrix}}.$$

Its eigenvalues and normalized eigenvectors are

$$\begin{array}{lcccc} \text{eigenvalues:} & 3e\mathscr{E}a_0 & -3e\mathscr{E}a_0 & \multicolumn{2}{c}{\overbrace{\qquad\qquad}^{0}} \\ \text{eigenvectors:} & \begin{pmatrix} 1/\sqrt{2} \\ 0 \\ -1/\sqrt{2} \\ 0 \end{pmatrix} & \begin{pmatrix} 1/\sqrt{2} \\ 0 \\ 1/\sqrt{2} \\ 0 \end{pmatrix} & \begin{pmatrix} 0 \\ 1 \\ 0 \\ 0 \end{pmatrix} & \begin{pmatrix} 0 \\ 0 \\ 0 \\ 1 \end{pmatrix}. \end{array}$$

Thus the degeneracy is partially lifted, the perturbed levels being $E_2 + 3e\mathscr{E}a_0$, $E_2 - 3e\mathscr{E}a_0$, and E_2; the last remains doubly degenerate. (As on p. 99), E_2 itself is $-\hbar^2/8ma_0^2$.)

In 1901, Voigt predicted that spectral lines should be split by an intense electric field, but decided that the effect, being (as he believed) second-order in $\mathscr{E}$, would be too small to see, using the available techniques. Normally this conclusion would be correct; but atomic hydrogen displays the very special degeneracy (E_{nl} is independent of l) which allows the effect to become *first*-order in $\mathscr{E}$ for $n \geqslant 2$. Of course, this occurred to no-one at the time, and it was not till 1913 that Johannes Stark observed the first-order effect which now carries his name. The visible lines he observed involved transitions from excited states with $n > 2$, and the splitting was therefore more complex.

The eigenstates of $\mathsf{H}_0 + \mathsf{H}'$ to first order in $\mathscr{E}$ are obtained by inspecting the eigenvectors of V; they are given here:

eigenvalue		eigenstate
$E_2 + 3e\mathscr{E}a_0$		$(\lvert 200\rangle - \lvert 210\rangle)/\sqrt{2}$
E_2	(degenerate)	$\begin{cases} \lvert 211\rangle \\ \lvert 21-1\rangle \end{cases}$
$E_2 - 3e\mathscr{E}a_0$		$(\lvert 200\rangle + \lvert 210\rangle)/\sqrt{2}$.

We shall use these eigenstates in Chapter 15.

PROBLEM

Consider the Stark effect for a charged particle constrained to move on a circle (see pp. 46, 47, and supplement the Hamiltonian with the perturbation $H' = -q\mathscr{E}a\cos\theta$). Show that the degeneracy is not lifted in first order.

The normal Zeeman effect

In 1896 Pieter Zeeman discovered that the lines in the prominent yellow doublet in the spectrum of atomic sodium were apparently *broadened* by the action of a strong magnetic field on the sodium atoms. More careful work soon showed that each line was in fact splitting into a number of components. It is to be expected that the effect comes about through the magnetic dipole moment of the atom, just as the Stark effect depends on the electric dipole moment. However, the details are very different. For the moment, we shall ignore the electron's *intrinsic* magnetic moment, for a reason to be explained later.

We begin by establishing an important connection between the magnetic moment of an (arbitrary) neutral atom and the total orbital angular momentum of its constituent electrons. There is a result of electromagnetic theory which says that a collection of particles with total zero charge, all moving in the vicinity of their centre of mass O, is approximately equivalent (when viewed from a long way off) to a magnetic dipole at O. If particle m_j has charge Q_j and position $\boldsymbol{r}_j$, the equivalent magnetic dipole is

$$\mathscr{M} = -\sum \tfrac{1}{2} Q_j \boldsymbol{r}_j \wedge \dot{\boldsymbol{r}}_j.$$

On the other hand, the total angular momentum about O is

$$\boldsymbol{M} = \sum_j m_j \boldsymbol{r}_j \wedge \dot{\boldsymbol{r}}_j.$$

Now, in a neutral atom, one particle (the nucleus) is very much more massive than the remainder, and its velocity is consequently small: small enough in fact for the corresponding terms in $\mathscr{M}$ and $\boldsymbol{M}$ to be negligible (we are effectively setting $\eta = 1$, p. 95). All the other particles have the same mass m_e and the same charge $-e$, and it immediately follows that

$$\mathscr{M} = \frac{e}{2m_e}\boldsymbol{M},$$

a very general classical result.

Since both angular momentum and magnetic moment are naturally additive entities, we may immediately deduce, by the correspondence principle, that the observable

$$\text{total magnetic moment } \mathscr{M} = \frac{e}{2m_e} \times (\text{total angular momentum } \boldsymbol{M}).$$

Now consider a hydrogen atom in a uniform magnetic field $\boldsymbol{B} = (0, 0, B_z)$. In the dipole approximation, $\boldsymbol{B}$ contributes a further perturbation term to the Hamiltonian

$$\mathsf{H}' = -\boldsymbol{B}.\mathcal{M} = -(B_z e/2m_e)\mathsf{m}_z.$$

The first-order perturbation calculation for the $n = 2$ states is now very simple, since each of the four states $|200\rangle$, $|211\rangle$, $|210\rangle$, $|21-1\rangle$ is an *eigenstate* of m_z with eigenvalues 0, $\hbar$, 0, $-\hbar$. The secular matrix is therefore

$$V = \begin{array}{c} (200) \\ (211) \\ (210) \\ (21-1) \end{array} \begin{array}{c} \begin{array}{cccc} (200) & (211) & (210) & (21-1) \end{array} \\ \begin{pmatrix} 0 & 0 & 0 & 0 \\ 0 & -\hbar B_z e/2m_e & 0 & 0 \\ 0 & 0 & 0 & 0 \\ 0 & 0 & 0 & +\hbar B_z e/2m_e \end{pmatrix} \end{array},$$

with the obvious eigenvalues $0, 0, \hbar B_z e/2m_e$, $-\hbar B_z e/2m_e$. Thus an external magnetic field will split the degenerate $n = 2$ level into three levels (one still degenerate), with consequential changes in the observed emission spectrum of the atom.

Unlike the Stark effect, the Zeeman effect remains first-order even in the absence of the l-degeneracy characteristic of the hydrogen atom. It can be shown that for *any* atom, any level with a non-zero total angular momentum l is l-fold degenerate (provided electron spin is ignored), and a magnetic field will cause a splitting into l components spaced by $\hbar Bq/2m_e$. On account of a *selection rule*, (p. 190), it turns out that the emitted spectrum is simpler than might be expected, since all the *observed* spectral lines are predicted to be *triplets*.

The anomalous Zeeman effect

Fortunately perhaps for the development of the theory, the relatively rare normal Zeeman effect was observed first. Only one year later, Thomas Preston observed some unexpected *quadruplets*, and it was not long before other observations showed that the Zeeman effect was much more complicated than was at first thought.

At the time, the normal Zeeman effect was easy to explain on *classical* lines; we have sketched the argument in the last section. In frequency terms, the level spacing is $Be/4\pi m_e$, and this could be predicted by some very general work by Larmor. In fact, Larmor's discussion was so general that no-one could conceive of the possibility of any effect other than the normal one. The explanation of the anomalous effect was not to be found for more than a quarter of a century, when the nature of electron spin began to be understood.

The fact is that the classical relation between angular momentum and magnetic moment is *false* for the intrinsic spin of the electron; the argument leading to the classical relation applies only to orbital angular momentum. For the electron, on the other hand,

$$\text{magnetic moment} = \frac{e}{m_e}\,(\text{spin})$$

almost exactly, with magnitude $\hbar e/2m_e$, anomalously large by a **Landé factor** of two. As a result, while

$$\text{total angular momentum} = \text{orbital a.m.} + \text{spin},$$

we have

$$\text{total magnetic moment} = \frac{e}{2m_e}\,(\text{orbital a.m.} + 2 \times \text{spin}),$$

and the simple connection which would give the normal effect is spoilt by the presence of the factor of two. A complete account of the anomalous Zeeman effect is outside of the scope of this book.

PROBLEMS

1. A hydrogen atom is placed simultaneously in a uniform electric field and a uniform magnetic field. The fields are parallel. Show that the secular matrix for the degenerate $n = 2$ level may be put in the form

$$\begin{pmatrix} 0 & 0 & \alpha & 0 \\ 0 & -\beta & 0 & 0 \\ \alpha & 0 & 0 & 0 \\ 0 & 0 & 0 & \beta \end{pmatrix},$$

where α and β are constants. Discuss the splitting of the level.

2. The fields of problem 1 are now arranged to be 'crossed' (at right angles). Show that the secular matrix may now be written

$$\begin{pmatrix} 0 & 2^{-\frac{1}{2}}\alpha & 0 & 2^{-\frac{1}{2}}\alpha \\ 2^{-\frac{1}{2}}\alpha & -\beta & 0 & 0 \\ 0 & 0 & 0 & 0 \\ 2^{-\frac{1}{2}}\alpha & 0 & 0 & \beta \end{pmatrix}.$$

Discuss the splitting of the level.

Hidden δ-functions

This section clears some ground for a study of hyperfine splitting in hydrogen.

If $r \neq 0$, then it is easy to show that

$$\nabla^2\left(\frac{1}{r}\right) = 0.$$

At $r = 0$, the left side is clearly singular, and the nature of this singularity has some important implications for quantum mechanics.

Integrate the left side through a sphere of radius a centred on O, and transform to a surface integral over the sphere by the standard method:

$$\iiint_{r<a} \mathrm{d}^3\boldsymbol{r}\nabla^2\left(\frac{1}{r}\right) = a^2 \iint \sin\theta\, \mathrm{d}\theta\, \mathrm{d}\phi\, \boldsymbol{n}_{\theta\phi} \cdot \left(\nabla\frac{1}{\mathrm{r}}\right)_{r=a}$$

where $\boldsymbol{n}_{\theta\phi}$ is the outward normal to the sphere. The right side is

$$a^2 \iint \sin\theta\, \mathrm{d}\theta\, \mathrm{d}\phi \left(\frac{\mathrm{d}}{\mathrm{d}r}\frac{1}{r}\right)_{r=a}$$

$$= -4\pi.$$

Thus

(i) $\nabla^2\left(\frac{1}{r}\right) = 0$ if $r \neq 0$

(ii) the integral of $\nabla^2\left(\frac{1}{r}\right)$ through a region containing O is -4π.

This can mean only one thing:

$$\nabla^2\left(\frac{1}{r}\right) = -4\pi\delta^3(\boldsymbol{r}).$$

By similar methods, we can obtain more detailed information:

$$\frac{\partial^2}{\partial x^2}\left(\frac{1}{r}\right) = -\frac{1}{r^3} + \frac{3x^2}{r^5} - \frac{4}{3}\pi\delta^3(\boldsymbol{r}), \text{ etc.,}$$

$$\frac{\partial^2}{\partial x \partial y}\left(\frac{1}{r}\right) = \frac{3xy}{r^5}, \text{ etc.}$$

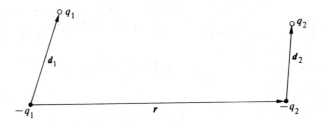

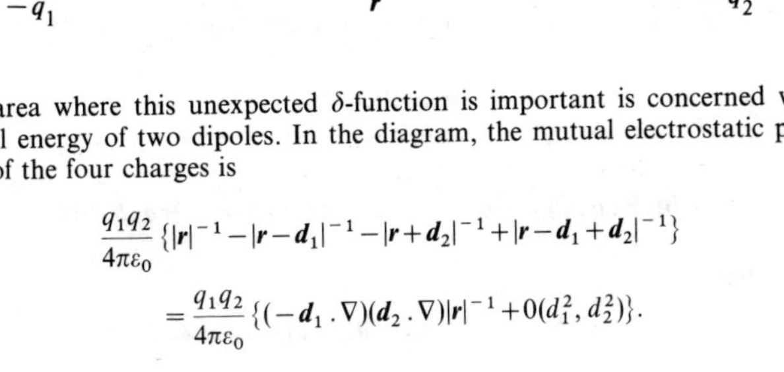

One area where this unexpected δ-function is important is concerned with the potential energy of two dipoles. In the diagram, the mutual electrostatic potential energy of the four charges is

$$\frac{q_1 q_2}{4\pi\varepsilon_0}\{|\boldsymbol{r}|^{-1} - |\boldsymbol{r}-\boldsymbol{d}_1|^{-1} - |\boldsymbol{r}+\boldsymbol{d}_2|^{-1} + |\boldsymbol{r}-\boldsymbol{d}_1+\boldsymbol{d}_2|^{-1}\}$$

$$= \frac{q_1 q_2}{4\pi\varepsilon_0}\{(-\boldsymbol{d}_1 \cdot \nabla)(\boldsymbol{d}_2 \cdot \nabla)|\boldsymbol{r}|^{-1} + 0(d_1^2, d_2^2)\}.$$

Let us now go to the dipole limit:

$$\boldsymbol{d}_1 \to 0,\ q_1 \to \infty, \text{ such that } q_1 \boldsymbol{d}_1 \to \boldsymbol{D}_1,$$

and similarly for $\boldsymbol{D}_2$; this gives

mutual P.E. of the two electric dipoles $\boldsymbol{D}_1$ and $\boldsymbol{D}_2$

$$= -\frac{1}{4\pi\varepsilon_0}(\boldsymbol{D}_1 . \nabla)(\boldsymbol{D}_2 . \nabla)|\boldsymbol{r}|^{-1}$$

where $\boldsymbol{r}$ is the vector separation of the dipoles.

There is a hidden δ-function here, which can be made explicit by carrying out the differentiations, using the results already given,

$$-(\boldsymbol{D}_1 . \nabla)(\boldsymbol{D}_2 . \nabla)\frac{1}{r} = r^{-3}\boldsymbol{D}_1 . \boldsymbol{D}_2 - 3r^{-5}(\boldsymbol{r} . \boldsymbol{D}_1)(\boldsymbol{r} . \boldsymbol{D}_2) + \tfrac{4}{3}\pi \boldsymbol{D}_1 . \boldsymbol{D}_2 \delta^3(\boldsymbol{r}).$$

Usually, in quoting the formula on the right, the δ-function term is dropped: who wants to know about two dipoles at *exactly* the same place? However, in quantum mechanics we may well wish to evaluate the *mean value* of such a potential energy, and this will involve an integration over all space. Clearly, the δ-function will then make an important contribution to the result. Physically, the new **contact term** may be viewed as providing a limiting correction when the dipoles are so close together as to discern each other's structure.

Surprisingly, magnetic dipoles behave differently. For the magnetic field (produced by currents, not charges) we must start with the vector potential produced by a dipole $\boldsymbol{m}_1$; it is

$$\boldsymbol{A}(\boldsymbol{r}) = \frac{\mu_0}{4\pi} \nabla \wedge (\boldsymbol{m}_1/r).$$

The magnetic field $\boldsymbol{B}$ is $\nabla \wedge \boldsymbol{A}$, and the potential energy of a second dipole in this field is $\boldsymbol{m}_2 . \boldsymbol{B}$. Thus the result we seek is

$$-\frac{\mu_0}{4\pi} \boldsymbol{m}_2 . \nabla \wedge \left\{\nabla \wedge \frac{\boldsymbol{m}_1}{r}\right\}$$

$$= \frac{\mu_0}{4\pi}\left\{(\boldsymbol{m}_1 . \boldsymbol{m}_2)\nabla^2 \frac{1}{r} - (\boldsymbol{m}_1 . \nabla)(\boldsymbol{m}_2 . \nabla)\frac{1}{r}\right\} \qquad \text{(see problem 2.)}$$

$$= \frac{\mu_0}{4\pi}\{r^{-3}\boldsymbol{m}_1 . \boldsymbol{m}_2 - 3r^{-5}(\boldsymbol{r} . \boldsymbol{m}_1)(\boldsymbol{r} . \boldsymbol{m}_2) - \tfrac{8}{3}\pi \boldsymbol{m}_1 . \boldsymbol{m}_2 \delta^3(\boldsymbol{r})\}.$$

In comparison with the electric dipole case, the contact term is twice as large with the opposite sign. This will soon be seen to be of paramount importance for the effects of mutual interaction of electron and nuclear magnetic moments in an atom.

PROBLEMS

1. Verify that

$$-(\boldsymbol{a} . \nabla)(\boldsymbol{b} . \nabla)\frac{1}{r} = r^{-3}\boldsymbol{a} . \boldsymbol{b} - 3r^{-5}(\boldsymbol{a} . \boldsymbol{r})(\boldsymbol{b} . \boldsymbol{r}) + \tfrac{4}{3}\pi \boldsymbol{a} . \boldsymbol{b}\delta^3(\boldsymbol{r}).$$

2. Show that

$$-\boldsymbol{a} . (\boldsymbol{b} \wedge (\boldsymbol{c} \wedge \boldsymbol{d})) = (\boldsymbol{a} . \boldsymbol{d})(\boldsymbol{b} . \boldsymbol{c}) - (\boldsymbol{a} . \boldsymbol{c})(\boldsymbol{b} . \boldsymbol{d}).$$

The hyperfine structure of the hydrogen atom ground state

Atomic nuclei possess magnetic moments, and are therefore also affected by magnetic fields. In particular they will interact with any field that happens to exist as a result of the structure of the atom itself, for example, the field of *electronic* magnetic moments. Consequently, once again degenerate atomic levels may be split, giving a **hyperfine structure** in the emission spectrum of the atom.

Consider the hydrogen atom (with the proton fixed at $\boldsymbol{r} = 0$, for convenience): the spin of both electron and proton are to be taken into account. We therefore need to set up an $\mathbf{r}\varepsilon_e\varepsilon_p$-representation (on the lines of the $\mathbf{r}\mathsf{m}_z$-representation, p. 125), in which

$$\text{for the electron: } (\mathsf{m}_z)_e = \tfrac{1}{2}\varepsilon_e\hbar \qquad (\varepsilon_e = \pm 1),$$
$$\text{for the proton: } (\mathsf{m}_z)_p = \tfrac{1}{2}\varepsilon_p\hbar \qquad (\varepsilon_p = \pm 1).$$

The wavefunction for a general state Ψ,

$$\psi_{\varepsilon_e\varepsilon_p}(\boldsymbol{r}) = \langle \boldsymbol{r}\varepsilon_e\varepsilon_p|\Psi\rangle,$$

has *four* components, conveniently arranged in a column

$$\begin{pmatrix} \psi_{++}(\boldsymbol{r}) \\ \psi_{+-}(\boldsymbol{r}) \\ \psi_{-+}(\boldsymbol{r}) \\ \psi_{--}(\boldsymbol{r}) \end{pmatrix}.$$

The spin observables will have to be represented by 4×4 matrices. Consider the x-component of electron spin, $\frac{1}{2}\hbar\sigma_{ex}$. By the comments on direct products (p. 125), this must act like a Pauli matrix on the suffix ε_e, and leave the suffix ε_p alone. Thus we require

$$\begin{pmatrix} \langle \boldsymbol{r}++| \\ \langle \boldsymbol{r}+-| \\ \langle \boldsymbol{r}-+| \\ \langle \boldsymbol{r}--| \end{pmatrix} \sigma_{ex} = \begin{pmatrix} \langle \boldsymbol{r}-+| \\ \langle \boldsymbol{r}--| \\ \langle \boldsymbol{r}++| \\ \langle \boldsymbol{r}+-| \end{pmatrix} = \begin{pmatrix} 0 & 0 & 1 & 0 \\ 0 & 0 & 0 & 1 \\ 1 & 0 & 0 & 0 \\ 0 & 1 & 0 & 0 \end{pmatrix} \begin{pmatrix} \langle \boldsymbol{r}++| \\ \langle \boldsymbol{r}+-| \\ \langle \boldsymbol{r}-+| \\ \langle \boldsymbol{r}--| \end{pmatrix}.$$

This shows us what the 4×4 replacement for the Pauli matrix σ_x for the electron must be. On the other hand, σ_x for the *proton* affects the suffix ε_p, leaving ε_e alone:

$$\begin{pmatrix} \langle \boldsymbol{r}++| \\ \langle \boldsymbol{r}+-| \\ \langle \boldsymbol{r}-+| \\ \langle \boldsymbol{r}--| \end{pmatrix} \sigma_{px} = \begin{pmatrix} \langle \boldsymbol{r}+-| \\ \langle \boldsymbol{r}++| \\ \langle \boldsymbol{r}--| \\ \langle \boldsymbol{r}-+| \end{pmatrix} = \begin{pmatrix} 0 & 1 & 0 & 0 \\ 1 & 0 & 0 & 0 \\ 0 & 0 & 0 & 1 \\ 0 & 0 & 1 & 0 \end{pmatrix} \begin{pmatrix} \langle \boldsymbol{r}++| \\ \langle \boldsymbol{r}+-| \\ \langle \boldsymbol{r}-+| \\ \langle \boldsymbol{r}--| \end{pmatrix}.$$

Hence we infer what the replacement of σ_x for the proton must be.

The complete set of six matrices turns out to be

$$\boldsymbol{\sigma}_e = \begin{pmatrix} 0 & 0 & 1 & 0 \\ 0 & 0 & 0 & 1 \\ 1 & 0 & 0 & 0 \\ 0 & 1 & 0 & 0 \end{pmatrix}, \begin{pmatrix} 0 & 0 & -i & 0 \\ 0 & 0 & 0 & -i \\ i & 0 & 0 & 0 \\ 0 & i & 0 & 0 \end{pmatrix}, \begin{pmatrix} 1 & 0 & 0 & 0 \\ 0 & 1 & 0 & 0 \\ 0 & 0 & -1 & 0 \\ 0 & 0 & 0 & -1 \end{pmatrix}$$

$$\boldsymbol{\sigma}_p = \begin{pmatrix} 0 & 1 & 0 & 0 \\ 1 & 0 & 0 & 0 \\ 0 & 0 & 0 & 1 \\ 0 & 0 & 1 & 0 \end{pmatrix}, \begin{pmatrix} 0 & -i & 0 & 0 \\ i & 0 & 0 & 0 \\ 0 & 0 & 0 & -i \\ 0 & 0 & i & 0 \end{pmatrix}, \begin{pmatrix} 1 & 0 & 0 & 0 \\ 0 & -1 & 0 & 0 \\ 0 & 0 & 1 & 0 \\ 0 & 0 & 0 & -1 \end{pmatrix}.$$

Having established the representation, we are now ready to apply the perturbation recipe. In the absence of any spin–spin interaction, the ground state has become *fourfold* degenerate, since its representative is

$$\psi_{\varepsilon_e\varepsilon_p}(\boldsymbol{r}) = \begin{pmatrix} c_{++} \\ c_{+-} \\ c_{-+} \\ c_{--} \end{pmatrix} \pi^{-\frac{1}{2}} a_0{}^{-\frac{3}{2}} \exp(-r/a_0) = v\pi^{-\frac{1}{2}} a_0{}^{-\frac{3}{2}} \exp(-r/a_0)$$

(p. 100), where *any* column vector v with constant coefficients will do. The perturbation is to be the interaction energy of the magnetic dipoles of the electron and proton, $m_e\boldsymbol{\sigma}_e$ and $m_p\boldsymbol{\sigma}_p$; this is

$$\mathsf{H}' = \frac{\mu_0}{4\pi} m_e m_p \{r^{-3}\boldsymbol{\sigma}_e \cdot \boldsymbol{\sigma}_p - 3r^{-5}(\boldsymbol{r} \cdot \boldsymbol{\sigma}_e)(\boldsymbol{r} \cdot \boldsymbol{\sigma}_p) - \tfrac{8}{3}\pi \boldsymbol{\sigma}_e \cdot \boldsymbol{\sigma}_p \delta^3(\boldsymbol{r})\}$$

(p. 168). This is a 4×4 matrix function of $\boldsymbol{r}$. The perturbation recipe calls for the evaluation of matrix elements like

$$\langle E_0 j|\mathsf{H}'|E_0 k\rangle = \pi^{-1} a_0^{-3} \int d^3\boldsymbol{r} \exp(-r/a_0) v_j^+ \mathsf{H}' v_k \exp(-r/a_0).$$

As it happens, on account of the spherical symmetry of the wavefunction, only the contact term in H' survives, and

$$\langle E_0 j|\mathsf{H}'|E_0 k\rangle = -\pi^{-1} a_0^{-3} \frac{\mu_0}{4\pi} m_e m_p \tfrac{8}{3}\pi v_j^+ (\boldsymbol{\sigma}_e \cdot \boldsymbol{\sigma}_p) v_k$$

$$= A v_j^+ (\boldsymbol{\sigma}_e \cdot \boldsymbol{\sigma}_p) v_k.$$

Using the table of constants on p. 242, with $m_p = +2{\cdot}793$ nuclear magnetons, $m_e = -1$ Bohr magneton, gives

$$A = 2{\cdot}354 \times 10^{-25}\ \text{J}.$$

The simplest orthonormal choice for the vectors v_j is clearly

$$\begin{pmatrix}1\\0\\0\\0\end{pmatrix}, \begin{pmatrix}0\\1\\0\\0\end{pmatrix}, \begin{pmatrix}0\\0\\1\\0\end{pmatrix}, \begin{pmatrix}0\\0\\0\\1\end{pmatrix},$$

giving as the 4×4 secular matrix

$$\begin{aligned} V &= A\boldsymbol{\sigma}_e \cdot \boldsymbol{\sigma}_p \\ &= A\{\sigma_{ex}\sigma_{px} + \sigma_{ey}\sigma_{py} + \sigma_{ez}\sigma_{pz}\} \\ &= A\begin{pmatrix} 1 & 0 & 0 & 0 \\ 0 & -1 & 2 & 0 \\ 0 & 2 & -1 & 0 \\ 0 & 0 & 0 & 1 \end{pmatrix}. \end{aligned}$$

The eigenvalues of V are A, A, A and $-3A$. Thus the fourfold degeneracy of the ground level is only partially lifted: there is a splitting into a degenerate triplet (shift A) and a singlet (shift $-3A$). The amount of the splitting is clearly $4A$.

In frequency terms, the gap is $4A/h = 1{\cdot}42 \times 10^9$ Hz. Compared with the frequencies associated with optical spectra, this is rather small, but the hyperfine splitting of optical lines can be observed if due care is taken. On the other hand, *microwave* transitions may be observed between the split ground-state levels themselves, yielding the celebrated **21 centimetre-line** of atomic hydrogen which plays such a large part in modern radio-astronomy.

The frequency of the hyperfine transition is one of the most accurately determined physical constants,

$$4A/h = 1\,420\,405\,751{\cdot}768 \pm 0{\cdot}001 \text{ Hz}.$$

(1979). Its precision is effectively limited by the precision of the standard atomic clock used to measure the period.

PROBLEMS

1. Show that

$$\int_{\text{all space}} \mathrm{d}^3\boldsymbol{r}\{f_1(r)\boldsymbol{a}\,.\,\boldsymbol{b} + f_2(r)(\boldsymbol{a}\,.\,\boldsymbol{r})(\boldsymbol{b}\,.\,\boldsymbol{r})\} = (\boldsymbol{a}\,.\,\boldsymbol{b})\int \mathrm{d}^3\boldsymbol{r}\{f_1(r) + \tfrac{1}{3}r^2 f_2(r)\}$$

where $\boldsymbol{a}$ and $\boldsymbol{b}$ are any two *constant* vectors. Deduce that in $\langle\Psi|\mathrm{H}'|\Psi\rangle$ only the contact term survives.

2. Obtain the eigenvalues and eigenvectors of V. Show that the singlet and triplet states have total angular momenta of 0 and $\hbar$ respectively. (First show that $(\boldsymbol{\sigma}_e + \boldsymbol{\sigma}_p)^2 = 6 + 2\boldsymbol{\sigma}_e \cdot \boldsymbol{\sigma}_p$.)

14. Time-dependent perturbation theory

Why do we need time-dependent perturbation theory?

PHYSICS is concerned with the way things happen in addition to the way things are. The last two chapters have been concerned with the changes experienced by stationary states as a consequence of a small perturbation; such changes are essentially *static* in character. Our concern now is with the ways in which non-stationary states may change in time, and this leads to a much more general kind of problem in which the Hamiltonian of the system may itself depend on time and in which, therefore, there may be no stationary states at all. Even when the Hamiltonian does *not* depend on time we may nevertheless wish to investigate non-stationary states; the mechanism by which an atomic system emits electromagnetic radiation is one of the most important examples of this.

Once again, perturbation theory is useful for those problems in which the Hamiltonian may be written as the sum of two terms

$$\mathsf{H}(t) = \mathsf{H}_0(t) + \mathsf{H}'(t),$$

the first of which may be dealt with exactly, and the second of which is in some sense small. Neither needs to be independent of the time, though in applications H_0 usually is, and H' sometimes is.

As usual, we shall derive a recipe of universal application. Because the workings of this recipe may be a little obscure at first, we shall introduce the method by a simple (non-quantum) example.

The method of variation of parameters

Consider the solution $x(t)$ of the ordinary differential equation

$$\dot{x} + x + \varepsilon x^2 = 0$$

for which $x(0) = x_0$ (given). Now, if $\varepsilon = 0$, the solution is immediate,

$$x = \mathrm{e}^{-t} x_0.$$

The method of *variation of parameters* begins by writing, for $\varepsilon \neq 0$,

$$x = \mathrm{e}^{-t} X(t)$$

as the solution of the original equation; the constant parameter x_0 has been replaced by the *varying* parameter $X(t)$. It is simple to verify that $X(t)$ must satisfy the equation

$$\dot{X} + \varepsilon\, \mathrm{e}^{-t} X^2 = 0, \quad \text{with} \quad X(0) = x_0.$$

The point is that, because $X(t)$ is constant when $\varepsilon = 0$, we expect $X(t)$ to vary *slowly* when ε is small; $X(t)$ is therefore a good candidate for expansion in powers of ε. So we write

$$X(t) = x_0 + \varepsilon\eta(t) + 0(\varepsilon^2)$$

and try to satisfy

$$\dot{X} + \varepsilon e^{-t}X^2 = \varepsilon\dot{\eta} + \varepsilon\, e^{-t}(x_0 + \varepsilon\eta)^2 = 0$$

to first order. Setting the coefficient of ε^1 to zero gives

$$\dot{\eta} + e^{-t}x_0^2 = 0$$

whence

$$\eta = x_0^2(e^{-t} - 1);$$

the constant of integration is chosen to ensure that $\eta(0) = 0$. The first-order solution of the original equation is thus

$$x(t) = e^{-t}\{x_0 + \varepsilon x_0^2(e^{-t} - 1) + 0(\varepsilon^2)\}.$$

Higher-order terms may be evaluated similarly, one by one.

By the way, the *exact* solution is easily verified:

$$x(t) = x_0/\{(1 + \varepsilon x_0)\, e^t - \varepsilon x_0\};$$

the expansion in powers of ε, of course, confirms the work just done. However, this expansion converges only if

$$|\varepsilon x_0(1 - e^{-t})| < 1,$$

and, no matter how small ε is, there are always values of t for which this condition is violated. This is a general feature of this kind of approximation, and we must therefore remember that, in each application, not only must ε be small, but t also must not be too large.

Solving the time-dependent Schrödinger equation approximately

We are given the general Schrödinger equation

$$i\hbar \frac{d}{dt}|\Psi t\rangle = \{\mathsf{H}_0(t) + \varepsilon\mathsf{H}'(t)\}|\Psi t\rangle,$$

and our aim is to expand the solution $|\Psi t\rangle$ as a power series in ε as a general-purpose recipe.

It is necessary to be able to solve the equation when $\varepsilon = 0$, somehow or other. How we do this is not important, but it is convenient to suppose for the moment that we have been able to find an operator $\mathsf{S}(t)$ satisfying the differential equation and initial condition

$$i\hbar \frac{d\mathsf{S}}{dt} = \mathsf{H}_0(t)\mathsf{S} \quad \text{and} \quad \mathsf{S}(0) = 1.$$

This is equivalent to the *complete* solution of the Schrödinger equation for

H_0 since, whatever the initial specification $|\Psi_0\rangle$,

$$i\hbar\frac{d}{dt}(S|\Psi_0\rangle) = H_0(t)(S|\Psi_0\rangle),$$

that is, $S(t)|\Psi_0\rangle$ is the solution in question. Actually to find the *explicit* form of $S(t)$ is often extremely inconvenient, but in all normal applications we shall find that there are ways of avoiding having to do this.

In the spirit of the method of the last section, we now write the sought solution as

$$|\Psi t\rangle = S(t)|\Phi t\rangle,$$

replacing the constant initial value $|\Psi_0\rangle$ by the (slowly) varying state vector $|\Phi t\rangle$ whose initial value is $|\Psi_0\rangle$.

Now we need to find the differential equation satisfied by $|\Phi t\rangle$. To this end, we write $i\hbar\, d|\Psi t\rangle/dt$ in two different ways. First, we have

$$\begin{aligned} i\hbar\frac{d}{dt}|\Psi t\rangle &= i\hbar\frac{d}{dt}\{S(t)|\Phi t\rangle\} \\ &= i\hbar\frac{dS}{dt}|\Phi t\rangle + Si\hbar\frac{d}{dt}|\Phi t\rangle \\ &= H_0(t)S|\Phi t\rangle + Si\hbar\frac{d}{dt}|\Phi t\rangle \\ &= H_0(t)|\Psi t\rangle + Si\hbar\frac{d}{dt}|\Phi t\rangle, \end{aligned}$$

where we have used the equation of motion for S. The second way of writing $i\hbar\, d|\Psi t\rangle/dt$ is given by the original Schrödinger equation of the problem. Comparing the two yields

$$S(t)i\hbar\frac{d}{dt}|\Phi t\rangle = \varepsilon H'(t)|\Psi t\rangle,$$

or

$$i\hbar\frac{d}{dt}|\Phi t\rangle = \varepsilon S^{-1}(t)H'(t)S(t)|\Phi t\rangle.$$

Thus $|\Phi t\rangle$ itself satisfies a Schrödinger equation, this time with an *inevitably* time-dependent Hamiltonian $\varepsilon S^{-1}H'S$. Up to this point, no approximations have been made.

The next step is to write

$$|\Phi t\rangle = |\Psi_0\rangle + \varepsilon|\Phi_1 t\rangle + 0(\varepsilon^2).$$

Substituting in the equation, and retaining only terms up to ε^1, gives

$$i\hbar \frac{d}{dt}|\Phi_1 t\rangle = \mathsf{S}^{-1}(t)\mathsf{H}'(t)\mathsf{S}(t)|\Psi_0\rangle.$$

Integrating each side from 0 to t gives

$$|\Phi_1 t\rangle = \frac{1}{i\hbar}\int_0^t d\tau \mathsf{S}^{-1}(\tau)\mathsf{H}'(\tau)\mathsf{S}(\tau)|\Psi_0\rangle$$

(note that $|\Phi_1, 0\rangle = 0$). Finally,

$$|\Psi t\rangle = \mathsf{S}(t)|\Phi t\rangle$$

$$= \mathsf{S}(t)\left\{1+\frac{\varepsilon}{i\hbar}\int_0^t d\tau \mathsf{S}^{-1}(\tau)\mathsf{H}'(\tau)\mathsf{S}(\tau)+0(\varepsilon^2)\right\}|\Psi_0\rangle.$$

Putting $\varepsilon = 1$, we have the general

Recipe: To solve the equation

$$i\hbar\frac{d}{dt}|\Psi t\rangle = \{\mathsf{H}_0(t)+\mathsf{H}'(t)\}|\Psi t\rangle, \quad |\Psi, t=0\rangle = |\Psi_0\rangle,$$

to first order in H':

(i) Find the operator $\mathsf{S}(t)$ satisfying

$$i\hbar\frac{d\mathsf{S}}{dt}(t) = \mathsf{H}_0(t)\mathsf{S}(t), \quad \mathsf{S}(0) = 1$$

(or some equivalent procedure);

(ii) Evaluate

$$|\Psi t\rangle \simeq \mathsf{S}(t)\left\{1+\frac{1}{i\hbar}\int_0^t d\tau \mathsf{S}^{-1}(\tau)\mathsf{H}'(\tau)\mathsf{S}(\tau)\right\}|\Psi_0\rangle.$$

The Stern–Gerlach experiment

As a first application of the recipe, we return to the Hamiltonian for the Stern–Gerlach experiment (pp. 74–6); in the $\boldsymbol{r}\sigma_z$-representation it is

$$H_0+H' = \left\{-\frac{\hbar^2}{2m}\nabla^2 - B_0\mu\sigma_z\right\}+\alpha\mu\begin{pmatrix} -z & -iy \\ iy & z \end{pmatrix}$$

after a slight rearrangement. The wavefunction $\psi(\boldsymbol{r}, t)$ is a position-dependent column vector with two components, whose initial value we shall take to be

$$\psi(\boldsymbol{r}, 0) = \begin{pmatrix} \chi_+ \\ \chi_- \end{pmatrix} e^{ikx}$$

representing a beam of particles with momentum $(\hbar k, 0, 0)$ and polarization $\begin{pmatrix}\chi_+\\ \chi_-\end{pmatrix}$. It is required to find $\psi(\boldsymbol{r}, t)$ to first order in α.

According to the recipe, we think first about $\mathsf{S}(t)$. When $\alpha = 0$, the equation is exactly satisfied by

$$\begin{pmatrix}\chi_+ \exp(-\tfrac{1}{2}\mathrm{i}\Omega t)\\ \chi_- \exp(\tfrac{1}{2}\mathrm{i}\Omega t)\end{pmatrix} \exp(\mathrm{i}(kx-\omega t))$$
$$= \exp(-\mathrm{i}\omega t)\begin{pmatrix}\exp(-\tfrac{1}{2}\mathrm{i}\Omega t) & 0\\ 0 & \exp(\tfrac{1}{2}\mathrm{i}\Omega t)\end{pmatrix}\psi(\boldsymbol{r}, 0)$$

(see pp. 75–6). Thus, *for our present purposes*, it is sufficient to write

$$\mathsf{S}(t) \leftarrow \exp(-\mathrm{i}\omega t)\begin{pmatrix}\exp(-\tfrac{1}{2}\mathrm{i}\Omega t) & 0\\ 0 & \exp(\tfrac{1}{2}\mathrm{i}\Omega t)\end{pmatrix},$$

(This expression describes the effect of $\mathsf{S}(t)$ exactly for initial conditions of the type we need to consider. This is good enough here. The specification would need to be augmented for a more general problem.)

Having found S, we may now write down the solution $\psi(\boldsymbol{r}, t)$ to first order in α. Evaluating the integral gives

$$\frac{1}{\mathrm{i}\hbar}\int_0^t \mathrm{d}\tau \mathsf{S}^{-1}(\tau)\mathsf{H}'\mathsf{S}(\tau)$$
$$\leftarrow \frac{1}{\mathrm{i}\hbar}\int_0^t \mathrm{d}\tau \exp(\mathrm{i}\omega t)\begin{pmatrix}\exp(\tfrac{1}{2}\mathrm{i}\Omega\tau) & 0\\ 0 & \exp(-\tfrac{1}{2}\mathrm{i}\Omega\tau)\end{pmatrix}.\alpha\mu\begin{pmatrix}-z & -\mathrm{i}y\\ \mathrm{i}y & z\end{pmatrix}$$
$$\times \exp(-\mathrm{i}\omega\tau)\begin{pmatrix}\exp(-\tfrac{1}{2}\mathrm{i}\Omega\tau) & 0\\ 0 & \exp(\tfrac{1}{2}\mathrm{i}\Omega\tau)\end{pmatrix}$$
$$= \frac{\alpha\mu}{\mathrm{i}\hbar}\int_0^t \mathrm{d}\tau\begin{pmatrix}-z & -\mathrm{i}y\exp(\mathrm{i}\Omega\tau)\\ \mathrm{i}y\exp(-\mathrm{i}\Omega\tau) & z\end{pmatrix}$$
$$= \frac{\alpha\mu}{\mathrm{i}\hbar}\begin{pmatrix}-zt & -y\,\Omega^{-1}(\exp(\mathrm{i}\Omega t)-1)\\ -y\,\Omega^{-1}(\exp(-\mathrm{i}\Omega t)-1) & zt\end{pmatrix}.$$

The solution correct to first order in α is therefore

$$\psi(\boldsymbol{r},t)\simeq\exp(-\mathrm{i}\omega t)\begin{pmatrix}\exp(-\frac{1}{2}\mathrm{i}\Omega t) & 0\\ 0 & \exp(\frac{1}{2}\mathrm{i}\Omega t)\end{pmatrix}$$

$$\times\left\{1+\frac{\alpha\mu}{\mathrm{i}\hbar}\begin{pmatrix}-zt & -y\,\Omega^{-1}(\exp(\mathrm{i}\Omega t)-1)\\ -y\,\Omega^{-1}(\exp(-\mathrm{i}\Omega t)-1) & zt\end{pmatrix}\right\}$$

$$\times\begin{pmatrix}\chi_+\\ \chi_-\end{pmatrix}\exp(\mathrm{i}kx)$$

as given by the recipe.

This completes the illustration of the method. However, one point needs to be made about experiments of this type: Ω is almost always very large. An electron in a field of 1 tesla has a precession frequency $\Omega/2\pi$ of about 3×10^{10} Hz. The time of one period, $2\pi/\Omega$, is expected to be much less than the typical time t spent between the pole pieces, and we may therefore ignore Ω^{-1} wherever it occurs, getting

$$\psi(\boldsymbol{r},t)\simeq\exp(-\mathrm{i}\omega t)\begin{pmatrix}\exp(-\frac{1}{2}\mathrm{i}\Omega t) & 0\\ 0 & \exp(\frac{1}{2}\mathrm{i}\Omega t)\end{pmatrix}$$

$$\times\begin{pmatrix}1-\alpha\mu zt/\mathrm{i}\hbar & 0\\ 0 & 1+\alpha\mu zt/\mathrm{i}\hbar\end{pmatrix}\begin{pmatrix}\chi_+\\ \chi_-\end{pmatrix}\exp(\mathrm{i}kx)$$

$$=\begin{pmatrix}\chi_+\exp(\mathrm{i}kx-\mathrm{i}(\omega+\frac{1}{2}\Omega)t+\mathrm{i}\alpha\mu zt/\hbar)\\ \chi_-\exp(\mathrm{i}kx-\mathrm{i}(\omega-\frac{1}{2}\Omega)t-\mathrm{i}\alpha\mu zt/\hbar)\end{pmatrix},$$

which is essentially the result at the bottom of page 76. (We have used the fact that $1+a\simeq e^a$ when a is small.) Note that ψ does not depend on y.

However, what if $\Omega=0$, as will happen if B_0 is zero on the x-axis? In this case

$$\Omega^{-1}(\exp(\mathrm{i}\Omega t)-1)\to\mathrm{i}t,$$

and

$$\psi(\boldsymbol{r},t)\simeq\exp(\mathrm{i}kx-\mathrm{i}\omega t)\begin{pmatrix}1-\mu\alpha zt/\mathrm{i}\hbar & -\mu\alpha yt/\hbar\\ \mu\alpha yt/\hbar & 1+\mu\alpha zt/\mathrm{i}\hbar\end{pmatrix}\begin{pmatrix}\chi_+\\ \chi_-\end{pmatrix}.$$

Since now ψ depends on y, the exit beam may be expected to have a more complicated structure. To see what this structure is, let us write

$$\boldsymbol{k}_\phi=\{k,(2\mu\alpha t/\hbar)\sin\phi,(2\mu\alpha t/\hbar)\cos\phi\},$$

and

$$f_\phi(\boldsymbol{r}, t) = \frac{1}{2\pi}\begin{pmatrix} 1+\cos\phi \\ -\mathrm{i}\sin\phi \end{pmatrix} \exp \mathrm{i}(\boldsymbol{k}_\phi . \boldsymbol{r} - \omega t)$$

$$= \frac{1}{2\pi}\begin{pmatrix} 1+\cos\phi \\ -\mathrm{i}\sin\phi \end{pmatrix} \exp \mathrm{i}(kx - \omega t)\left\{1 - \frac{2\mu\alpha t}{\hbar}(y\sin\phi + z\cos\phi)\right\}$$

$$+0(\alpha^2).$$

This function represents a beam of particles with momentum k_ϕ, polarized in the direction $(0, \sin\phi, \cos\phi)$; the momenta for different values of ϕ all lie in the surface of a circular cone of semiangle $2\mu\alpha t/k\hbar$. By integrating the last expression for f_ϕ, it is straightforward to show that

$$\int_0^{2\pi} \mathrm{d}\phi f_\phi(\boldsymbol{r}, t) = \exp(\mathrm{i}kx - \mathrm{i}\omega t)\begin{pmatrix} 1+\mathrm{i}\mu\alpha z t/\hbar \\ \mu\alpha y t/\hbar \end{pmatrix},$$

and this is just $\psi(\boldsymbol{r}, t)$ in the case $\chi_+ = 1$, $\chi = 0$. It follows that the exit beam does not have two components, but is spread over the surface of a *cone*.

An analogous optical phenomenon may be observed when light is refracted by a biaxial crystal: a beam in the correct direction of incidence is refracted into a conical shape. There are important differences of detail, arising from the fact that photons and electrons carry angular momenta of $\hbar$ and $\frac{1}{2}\hbar$ respectively. In the Stern–Gerlach case it would not be difficult in principle to generate an inhomogeneous field of the right kind—four knife-edge pole pieces will do—but the practical difficulties are likely to be severe, and as far as I know the experiment has never been performed.

Transition probabilities

Before 1925, in the days of the 'older' quantum theory, atomic systems were supposed to emit radiation by making **transitions**: the constituent electrons were supposed to make sudden inexplicable jumps from one state of motion to another. This idea is no longer appropriate in modern theory, but the name has stuck: if the energy of a quantum system is observed to be E_i to begin with, and then, on account of the effect of some perturbation, is later observed to be E_f, we say—improperly—that the system has made a transition from the initial state $|\boldsymbol{i}\rangle$ (of energy E_i) to an orthogonal final state $|\boldsymbol{f}\rangle$ (of energy E_f). An important application of the recipe (p. 175) is to the evaluation of the **transition probability** $P(\boldsymbol{i} \to \boldsymbol{f})$, which by the usual rule is

$$P(\boldsymbol{i} \to \boldsymbol{f}) = |\langle \boldsymbol{f}|\Psi t\rangle|^2, \quad \text{when} \quad |\Psi 0\rangle = |\boldsymbol{i}\rangle.$$

Here, both $|\boldsymbol{i}\rangle$ and $|\boldsymbol{f}\rangle$ are eigenstates of the *constant* H_0, with isolated eigenvalues E_i and E_f.

First we consider $\mathsf{S}(t)$. For any (time-dependent) eigenstate of H_0 we require

$$|Et\rangle = \mathsf{S}(t)|E0\rangle,$$

and we know

$$|Et\rangle = |E0\rangle \exp(-\text{i}Et/\hbar).$$

Thus

$$\mathsf{S}(t)|E0\rangle = |E0\rangle \exp(-\text{i}Et/\hbar).$$

It follows that

$$\mathsf{S}^{-1}(t)|E0\rangle = |E0\rangle \exp(\text{i}Et/\hbar).$$

This is all we need to know about $\mathsf{S}(t)$ in this instance.

Now we can evaluate (to first order)

$$\begin{aligned}\langle \boldsymbol{f}|\Psi t\rangle &\simeq \langle \boldsymbol{f}|\mathsf{S}(t)\left\{1+\frac{1}{\text{i}\hbar}\int_0^t \text{d}\tau \mathsf{S}^{-1}(\tau)\mathsf{H}'(\tau)\mathsf{S}(\tau)\right\}|\boldsymbol{i}\rangle \\ &= \exp(-\text{i}E_\text{f}t/\hbar)\langle \boldsymbol{f}|\boldsymbol{i}\rangle \\ &\quad +\frac{1}{\text{i}\hbar}\exp(-\text{i}E_\text{f}t/\hbar)\int_0^t \text{d}\tau \exp(\text{i}E_\text{f}\tau/\hbar)\langle \boldsymbol{f}|\mathsf{H}'(\tau)|\boldsymbol{i}\rangle \exp(-\text{i}E_\text{i}\tau/\hbar).\end{aligned}$$

The first term is zero, by orthogonality. The second term is obtained by moving $\langle \boldsymbol{f}|$ and $|\boldsymbol{i}\rangle$ until only H' separates them, using the properties of S. Thus we have immediately, to this order in H',

$$P(\boldsymbol{i}\to\boldsymbol{f}) = \frac{1}{\hbar^2}\left|\int_0^t \text{d}\tau\langle \boldsymbol{f}|\mathsf{H}'(\tau)|\boldsymbol{i}\rangle \exp(\text{i}(E_\text{f}-E_\text{i})\tau/\hbar)\right|^2,$$

a formula of wide application.

Resonance

It is important to notice that the integral is nearly—the limits of integration are wrong—the Fourier component of $\langle \boldsymbol{f}|\mathsf{H}'(\tau)|\boldsymbol{i}\rangle$ with frequency $(E_\text{i}-E_\text{f})/\hbar$. It appears that only those frequency components which *match* the gap between the energy levels E_i and E_f will contribute significantly to the transition probability. This is the phenomenon of **resonance**, with far-reaching consequences for modern experimental physics; it is the cornerstone of the theory of laser action. It also provides us with our first rigorous explanation of the photoelectric effect, which is where we came in (p. 2). If before the ejection of an electron the energy of a block of metal is E_i, and after the ejection the energy of the block and electron together is

E_f, then the ejection will be achieved only if the incident radiation has a component of frequency $\nu = (E_f - E_i)/h$. Thus $h\nu$ must be at least as great as the kinetic energy of the ejected electron—even greater, if a potential step at the surface of the block is to be surmounted.

The approximate formula of the last section gives a transition probability proportional to t^2 for small t. For some purposes this is not quite good enough, and we shall now discuss a different kind of approximation which is appropriate in cases of resonance. The starting-point is the variation of parameters, but the method is not a perturbation one in the usual sense.

Consider the operator $\mathsf{S}^{-1}(t)\mathsf{H}'(t)\mathsf{S}(t)$ which occurs in the equation for $|\Phi t\rangle$. When H' contains an oscillating component $\exp(-\mathrm{i}\omega t)$ whose frequency satisfies the resonance condition $\hbar\omega = E_m - E_n$, then, because S^{-1}, H', S respectively contain oscillating contributions $\exp(\mathrm{i}E_m t/\hbar)$, $\exp(-\mathrm{i}\omega t)$, and $\exp(-\mathrm{i}E_n t/\hbar)$, their product $\mathsf{S}^{-1}\mathsf{H}'\mathsf{S}$ must contain a *constant* contribution—in addition to other oscillating ones. The **resonance approximation** consists in *throwing away all the oscillating contributions*, on the grounds that they cannot produce more than a small jitter in the solution $|\Phi t\rangle$ anyway.

The simplest and most frequent case is when H' contains contributions $\exp(\pm\mathrm{i}\omega t)$ of one frequency ω only, and H_0 contains *just two* levels E_1 and E_2 satisfying the condition $E_1 - E_2 = \hbar\omega$. Since the gaps between other levels do not satisfy the condition, we may throw these other levels away, and we are then left with an **effective Hamiltonian** which is a 2×2 matrix with elements $\langle E_m|\mathsf{S}^{-1}\mathsf{H}'\mathsf{S}|E_n\rangle$ $(m, n = 1 \text{ or } 2)$ that is

$$\mathsf{S}^{-1}\mathsf{H}'\mathsf{S} \leftarrow$$

$$\begin{pmatrix} \exp(\mathrm{i}E_1t/\hbar)\langle E_1|\mathsf{H}'|E_1\rangle\exp(-\mathrm{i}E_1t/\hbar) & \exp(\mathrm{i}E_1t/\hbar)\langle E_1|\mathsf{H}'|E_2\rangle\exp(-\mathrm{i}E_2t/\hbar) \\ \exp(\mathrm{i}E_2t/\hbar)\langle E_2|\mathsf{H}'|E_1\rangle\exp(-\mathrm{i}E_1t/\hbar) & \exp(\mathrm{i}E_2t/\hbar)\langle E_2|\mathsf{H}'|E_2\rangle\exp(-\mathrm{i}E_2t/\hbar) \end{pmatrix}$$

Now, we may write

$$\mathsf{H}' = \mathsf{A}\exp(-\mathrm{i}\omega t) + \mathsf{A}^+\exp(\mathrm{i}\omega t),$$

where A is a constant operator, not necessarily Hermitian. Inserting this in the matrix and throwing away all the remaining oscillatory terms leaves the constant residue

$$\mathsf{S}^{-1}\mathsf{H}'\mathsf{S} \leftarrow \begin{pmatrix} 0 & \langle E_1|\mathsf{A}|E_2\rangle \\ \langle E_2|\mathsf{A}^+|E_1\rangle & 0 \end{pmatrix}$$

$$= \begin{pmatrix} 0 & W \\ W^* & 0 \end{pmatrix}$$

say.

The solution for $|\Phi t\rangle$ is very simple in this approximation. Writing $c_i(t) = \langle E_i|\Phi t\rangle$ leads to

$$i\hbar\frac{\mathrm{d}}{\mathrm{d}t}\begin{pmatrix}c_1\\ c_2\end{pmatrix}=\begin{pmatrix}0 & W\\ W^* & 0\end{pmatrix}\begin{pmatrix}c_1\\ c_2\end{pmatrix},$$

showing that c_1 and c_2 must execute simple harmonic motion of angular frequency $\Omega = |W|/\hbar$. The solution of greatest interest is the one for which $c_1(0) = 1$, $c_2(0) = 0$: the system starts off in the state $|E_1\rangle$; this solution is

$$\begin{pmatrix}\cos\Omega t\\ \gamma\sin\Omega t\end{pmatrix},\quad \text{with}\quad \gamma=\frac{|W|}{\mathrm{i}W}.$$

Thus the state of the system swings steadily from $|E_1\rangle$ to $|E_2\rangle$ and back; the time needed to switch from either state to the other is $T = \pi/2\Omega = \pi\hbar/2|W|$.

Resonance is very familiar in classical physics. Sometimes it is regarded as desirable, as in the theory of tuned circuits, sometimes quite the opposite, as in civil and mechanical engineering. In acoustics, it depends on the circumstances. In quantum mechanics, most (not all) phenomena of any interest depend on resonance of one kind or another. In all these cases the mathematical approach is *essentially* the same: to use variation of parameters and ignore all oscillating terms.

PROBLEMS

1. Show that the fact

$$|\gamma\sin\Omega t|^2 = \Omega^2 t^2 + 0(t^4)$$

is in accord with the perturbation formula for $P(i\to f)$ of the last section.

2. An unperturbed system has three levels E_1, E_2, E_3 for which $E_1 - E_2 = E_2 - E_3 = \hbar\omega$, and is influenced by

$$\mathsf{H}' = \mathsf{A}\exp(-\mathrm{i}\omega t) + \mathsf{A}^+\exp(\mathrm{i}\omega t).$$

Show that this leads to an effective Hamiltonian of the form

$$\begin{pmatrix}0 & W_1 & 0\\ W_1^* & 0 & W_2\\ 0 & W_2^* & 0\end{pmatrix}.$$

Show that the initial state is revisited at intervals of $\pi\hbar/\{|W_1|^2 + |W_2|^2\}^{\frac{1}{2}}$.

Fermi's Golden Rule

Very frequently it happens that we are interested in a transition from a discrete state $\boldsymbol{i}$ to a state $\boldsymbol{f}$ which is *not* discrete, but which rather belongs to a continuous band of states. We then need to carry the work a little further;

it is worth doing this in a general way, since we obtain a simple formula of such wide application that in 1950 Fermi called it the **Golden Rule**.

To fix ideas, think of two quantum systems A and B, to be regarded for the moment as completely independent. System A is to have—among possible others—two isolated energy eigenstates with eigenvalues E_0 and E_1; system B has—again among possible others—an isolated eigenstate (ε_0), and a *continuous range* of eigenstates (ε_k), indexed with a continuous parameter k. Now imagine A and B to be *coupled* by a small perturbation

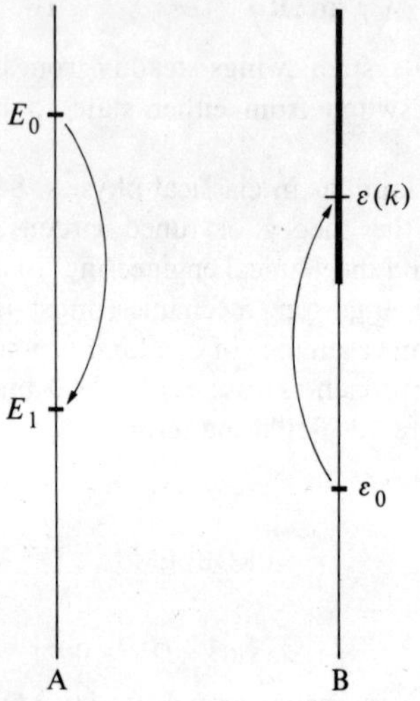

H′ which can cause transitions; H′ will always be assumed to be a *constant* operator for convenience in developing the theory—in any case, this is almost always true in practice.

The particular transition which interests us is $\boldsymbol{i \to f}$, where $\boldsymbol{i}$ and $\boldsymbol{f}$ are certain states of the composite system A+B, as follows:

composite state	notation	state of system A	state of system B
i	$\lvert E_0, \varepsilon_0\rangle$	E_0	ε_0
f	$\lvert E_1, k\rangle$	E_1	ε_k

Thus the transition $\boldsymbol{i \to f}$ means that system A makes a transition from E_0 to E_1, while B simultaneously makes a transition from ε_0 to one or other of the continuous range ε_k. Of course, if the coupling is small, we expect that energy conservation will require that ε_k is, at least approximately, $E_0+\varepsilon_0-E_1$. However, such a conclusion must be approximate only, since the very fact that transitions may now occur implies that states that were previously stationary are no longer stationary, and are no longer strict eigenstates of the Hamiltonian.

Before we turn to finding the general formula, let us consider two of its possible applications. First suppose that an atom (system A) in an excited state E_0 is surrounded by complete darkness. Such a system is weakly coupled to the electromagnetic field (system B), and we therefore expect that the atom may make a **spontaneous transition** to a lower state E_1, while the field makes a simultaneous upward transition from darkness to another state where there is one photon present with energy $\varepsilon \sim E_0-E_1$. This is precisely the mechanism behind the ordinary 'emission' of a photon by an excited atom, now described in a rather different way.

A second application is to the phenomenon of **autoionization**. Suppose now that A and B are two different electrons of the same atom or molecule, and that each is in an excited level, respectively E_0 and ε_0. If electron A makes a transition to a *lower* level E_1, enough energy may become available to enable electron B to leave the atom altogether; the condition for this to happen is that $\varepsilon \sim \varepsilon_0+E_0-E_1$ should lie in the continuous band of free states. This is the basis of the **Auger effect**, first observed in 1926: a high-energy photon ejects an electron from the inner shell of a heavy atom; a second electron falls from an outer shell to the inner shell; in doing so it makes available enough energy to allow yet another electron to escape. *Two* electrons are ejected in total, and the phenomenon is sometimes called the *double photoelectric effect*.

Now for the details. Since the final state $|E_1, k\rangle$ belongs to a continuous range of states the transition probability to *precisely* this state has little meaning. What we must do is to calculate the overall transition probability to some state f or other lying in some *range* of k-values; this implies an integration over this range. As usual, we must normalize the final state according to

$$\langle E_1, k_1|E_1, k_2\rangle = \delta(k_1-k_2).$$

With this normalization,

$$P(\boldsymbol{i}\to f \text{ in some range})$$
$$= \int_{\text{range}} \mathrm{d}k \frac{1}{\hbar^2}\left|\int_0^t \mathrm{d}\tau\langle E_1, k|\mathsf{H}'|E_0, \varepsilon_0\rangle \exp\left(\mathrm{i}(E_\mathrm{f}-E_\mathrm{i})\tau/\hbar\right)\right|^2.$$

where now $E_\mathrm{i} = E_0+\varepsilon_0$ and $E_\mathrm{f} = E_1+\varepsilon_k$. Since we are here assuming H' to be a *constant* operator, the integral over τ is easily evaluated, giving

$$P = \int_{\text{range}} \mathrm{d}k \frac{1}{\hbar^2} \left| \langle E_1, k|\mathrm{H}'|E_0, \varepsilon_0\rangle \frac{\exp(\mathrm{i}(E_f - E_i)t/\hbar) - 1}{\mathrm{i}(E_f - E_i)/\hbar} \right|^2.$$

$$= 2 \int_{\text{range}} \mathrm{d}k \frac{1}{\hbar^2} |\langle E_1, k|\mathrm{H}'|E_0, \varepsilon_0\rangle|^2 \frac{1 - \cos\{(\varepsilon(k) + E_1 - \varepsilon_0 - E_0)t/\hbar\}}{\{\varepsilon(k) + E_1 - \varepsilon_0 - E_0\}^2/\hbar^2}.$$

Up to this point, we have made no approximations other than working to the lowest order of smallness in H'.

The Golden Rule is the result of a compromise which is very often justified. If the time t is not too long, we expect that our formula will give a good estimate for P. On the other hand, if t is long enough, the last factor of the integrand is substantially equivalent to a δ-function: for, since for large α

$$(1 - \cos \alpha x)/\pi\alpha x^2 \sim \delta(x)$$

(see p. 115), we deduce that the last factor of the integrand may be replaced with

$$\pi t \delta(\{\varepsilon(k) + E_1 - \varepsilon_0 - E_0\}/\hbar),$$

provided t is large enough. Finding a suitable value of t is often possible, and is particularly easy in the case of atomic spectra.

So let us replace the last factor of the integrand by the δ-function and select our range of k values to include k_1 for which

$$\varepsilon(k_1) = E_0 + \varepsilon_0 - E_1,$$

that is, the value of k for which *strict* energy conservation holds. On account of the δ-function, the integral over k becomes elementary, and gives **Fermi's Golden Rule**:

$$P(\boldsymbol{i} \to \boldsymbol{f}) = \frac{2\pi t}{\hbar} \rho(k_1) |\langle E_1, k_1|\mathrm{H}'|E_0, \varepsilon_0\rangle|^2$$

where the factor ρ is given by

$$\frac{1}{\rho(k_1)} = \frac{\mathrm{d}\varepsilon(k_1)}{\mathrm{d}k_1},$$

and arises from a change of variable from k to $\varepsilon(k)$ (see problem 1, p. 115).

The Golden Rule works well when the perturbation is such as not to disturb energy conservation overmuch. In atomic transitions, the frequency condition

$$h\nu = E_0 - E_1$$

is not strictly satisfied observationally: emitted photons have energies which are spread over a narrow range centred on $h\nu$. In other words, spectral lines are observed to have a finite **natural width**. However, the observed width is of the order of 1 part in at least 10^6 in the visible spectrum, and the use of the Golden Rule is therefore amply justified. Since P is proportional to t, the constant of proportionality is the rate of spontaneous emission, namely,

$$\frac{2\pi}{\hbar}\rho(k_1)|\langle E_1, k_1|\mathsf{H}'|E_0, \varepsilon_0\rangle|^2 \text{ photons per excited atom per second.}$$

A method of calculating this emission rate is given in chapter 15.

The restriction to a composite of two systems A and B, described at the beginning of this section, is quite unnecessary. All that the Golden Rule requires is that the final state should belong to a continuous range.

15. Electric dipole radiation

Dipole radiation

'RADIATION' is a big subject. This chapter will cover only a tiny part: the emission and absorption of photons by small atomic or molecular systems.

Any system which includes electric charge and current is capable of absorbing and, in certain circumstances, emitting electromagnetic radiation. The mechanisms may be quite complicated, but for systems which are compact in comparison to the radiation wavelength it is possible to analyse the radiation into a sequence of contributions of (usually) decreasing importance: electric dipole, magnetic dipole, electric quadrupole, ... (Actually, these correspond to terms in the expansion of the radiation field into spherical harmonics.) In the majority of cases, the electric dipole mechanism is by far the most important, and will be the only kind considered here. However, the other mechanisms must not be overlooked; the emission of the hydrogen 21 cm-line (p. 169) proceeds by way of the magnetic dipole, for example.

Any system—an atom or a molecule, say—with an electric dipole moment $\boldsymbol{D}$ will interact with an electric field $\mathscr{E}$; as we have already seen in the Stark effect (p. 162), the interaction energy is $-\boldsymbol{D}.\mathscr{E}$. Of course, the Stark effect was a static affair, and was handled with eigenvalue perturbation methods. Now we have to deal with time-dependent effects, and the methods of the last chapter are more appropriate.

We expect that a first-order perturbation treatment will give very good results for this reason: the diameter of a radiating atom or molecule is usually at least three orders of magnitude smaller than the wavelength of the emitted radiation (as far as *optical* wavelengths are concerned; the disparity is less for X-rays). Considered as an 'antenna', the atom is very inefficient (efficient antennae are at least comparable in size to the wavelength of the associated radiation). In other words, the coupling is weak, and the perturbation approach should work well.

Spontaneous emission

A possible use for Fermi's Golden Rule has been suggested already (p. 183): an excited atom (E_1) in complete darkness emits a photon ($\hbar\omega$), and in doing so makes a transition to a lower level (E_0). The calculation falls into two parts: the first establishes a general relation between the emission rate and a matrix element of the electric dipole moment, and straightforwardly disposes of the radiative aspects of the problem once and for all; this is the objective of this section. It will then remain to work out the relevant dipole matrix elements for the particular atom (or molecule) under consideration; this may not be a trivial task for any but the simplest of systems.

The initial state $\boldsymbol{i}$ is $|E_1; \text{dark}\rangle$, and the final state $\boldsymbol{f}$ belongs to the continuous range of states $|E_0; \boldsymbol{k}, \boldsymbol{\lambda}\rangle$; the momentum $\hbar\boldsymbol{k}$ and the polarization $\boldsymbol{\lambda}$ of the emitted photon are discussed on pp. 145–7. The interaction energy is

$$\mathsf{H}' = -\mathbf{D}\,.\,\mathscr{E}(0),$$

where $\mathbf{D}$ is the electric dipole moment observable of the atom, and $\mathscr{E}(0)$ is the electric field intensity at the atom—at the origin, for convenience.

We are ready to apply the Golden Rule (p. 184). First, we evaluate, at time $t = 0$,

$$\begin{aligned}\langle E_0; \boldsymbol{k}, \boldsymbol{\lambda}|\mathsf{H}'|E_1; \text{dark}\rangle &= -\langle E_0|\mathbf{D}|E_1\rangle\,.\,\langle \boldsymbol{k}, \boldsymbol{\lambda}|\mathscr{E}(0)|\text{dark}\rangle \\ &= -\,\boldsymbol{d}\,.\,\boldsymbol{\lambda}^*(\hbar\omega/16\pi^3\varepsilon_0)^{\frac{1}{2}} \quad \text{with} \quad \omega = c|\boldsymbol{k}|\end{aligned}$$

(see the comment on the direct product, p. 125, and the matrix element of $\mathscr{E}$, p. 147; for convenience, we write $\boldsymbol{d} = \langle E_0|\mathbf{D}|E_1\rangle$.) The Golden Rule now gives the transition rate for $\boldsymbol{i} \to \boldsymbol{f}$

$$\frac{2\pi}{\hbar}\int \mathrm{d}^3\boldsymbol{k}\,\delta(\hbar\omega_0 - \hbar\omega) \sum_{1,\,2} |\boldsymbol{d}\,.\,\boldsymbol{\lambda}_i^*|^2(\hbar\omega/16\pi^3\varepsilon_0),$$

where $\hbar\omega_0 = E_1 - E_0$, the expected energy for the emitted photon. The sum over two orthogonal polarization states $\boldsymbol{\lambda}_1$ and $\boldsymbol{\lambda}_2$ for each $\boldsymbol{k}$, followed by the integral over all $\boldsymbol{k}$, takes care of all the one-photon states, while the δ-function picks out only those for which energy is conserved.

The integral may be evaluated exactly. First, we deal with the sum. The squared 'length' of the (possibly complex) vector $\boldsymbol{d}$ may be written in terms of components,

$$\boldsymbol{d}\,.\,\boldsymbol{d}^* = |\boldsymbol{d}\,.\,\boldsymbol{\lambda}_1^*|^2 + |\boldsymbol{d}\,.\,\boldsymbol{\lambda}_2^*|^2 + |\boldsymbol{d}\,.\,\boldsymbol{u}|^2,$$

where $\boldsymbol{u} = \boldsymbol{k}k^{-1}$ is the unit vector in the direction of $\boldsymbol{k}$. Thus

$$\sum_{1,\,2} |\boldsymbol{d}\,.\,\boldsymbol{\lambda}_i^*|^2 = \boldsymbol{d}\,.\,\boldsymbol{d}^* - (\boldsymbol{d}\,.\,\boldsymbol{u})(\boldsymbol{d}^*\,.\,\boldsymbol{u}).$$

Next, we write $\mathrm{d}^3\boldsymbol{k} = k^2\,\mathrm{d}k\,\mathrm{d}^2\Omega$, that is, the integral over all $\boldsymbol{k} = k\boldsymbol{u}$ may be considered as an integral over all positive k and all 'directions' $\boldsymbol{u}$; this last is just the integral over all solid angle Ω. Thus the

$$\begin{aligned}\text{transition rate} &= \frac{2\pi}{\hbar}\cdot\frac{\hbar\omega_0}{16\pi^3\varepsilon_0}\int_0^\infty k^2\,dk\delta(\hbar\omega_0-\hbar\omega)\\ &\quad\times\int d^2\Omega\,\{\boldsymbol{d}.\boldsymbol{d}^*-(\boldsymbol{d}.\boldsymbol{u})(\boldsymbol{d}^*.\boldsymbol{u})\}\\ &= \frac{\omega_0^3}{8\pi^2\varepsilon_0\hbar c^3}\int d^2\Omega\{\boldsymbol{d}.\boldsymbol{d}^*-(\boldsymbol{d}.\boldsymbol{u})(\boldsymbol{d}^*.\boldsymbol{u})\}.\end{aligned}$$

Finally, we may replace the second term in the integrand with $-\frac{1}{3}\boldsymbol{d}.\boldsymbol{d}^*$ (compare problem 1, p. 171), to give the

$$\text{transition rate} = \frac{\omega_0^3}{3\pi\varepsilon_0\hbar c^3}\boldsymbol{d}.\boldsymbol{d}^* \text{ photons per excited atom per second.}$$

This result is *completely general*. In any particular case, everything now depends on evaluating the matrix element of the dipole moment

$$\boldsymbol{d} = \langle E_0|\mathbf{D}|E_1\rangle.$$

The *power* radiated by a collection of N atoms is immediate when we recall that the energy per photon is $\hbar\omega_0$; in fact,

$$\begin{aligned}\text{mean radiated power} &= N.\hbar\omega_0.\text{transition rate}\\ &= \frac{N\omega_0^4}{3\pi\varepsilon_0 c^3}\boldsymbol{d}.\boldsymbol{d}^*.\end{aligned}$$

This formula enables us to calculate the *relative intensities* of spectral lines from a knowledge of the dipole moment matrix elements.

Notice, by the way, that $\hbar$ does not appear in the expression for the radiated power, suggesting that there must be a classical counterpart. In fact, a classical oscillating dipole $\boldsymbol{d}_c \cos\omega_0 t$ radiates a total power

$$\frac{\omega_0^4}{3\pi\varepsilon_0 c^3}\langle d_c^2\cos^2\ \omega_0 t\rangle_{\text{time average}}$$

and the similarity in the expressions has led some authors to favour a hybrid **semiclassical radiation theory**. My own reaction to such an approach is that it tends to introduce classical and quantum concepts in a rather unsatisfactory mixture, with a slight simplification in the mathematical development as the only compensation: it is best avoided, and is not described here.

PROBLEMS

1. Show that (a) $\int_0^\infty k^2\,\mathrm{d}k\delta(\hbar\omega_0-\hbar\omega)=\omega_0^2/\hbar c^3$,

   (b) $\int \mathrm{d}^2\Omega\{\boldsymbol{d}.\boldsymbol{d}^*-(\boldsymbol{d}.\boldsymbol{u})(\boldsymbol{d}^*.\boldsymbol{u})\}=(8\pi/3)\boldsymbol{d}.\boldsymbol{d}^*$.

2. For the hydrogen atom, the dipole moment $\boldsymbol{D}=-e\mathbf{r}$. Show that

$$\langle 100|\mathbf{D}|210\rangle = -e\,a_0\int \mathrm{d}^3\boldsymbol{\rho}\,.\,\pi^{-\frac{1}{2}}\,\mathrm{e}^{-\rho}\,.\,\boldsymbol{\rho}\,.\,(32\pi)^{-\frac{1}{2}}\rho\,\mathrm{e}^{-\rho/2}\cos\theta$$

$$= -0{\cdot}745\,e\,a_0(0,0,1),$$

and that $\omega_0=1{\cdot}55\times10^{16}\,\mathrm{s}^{-1}$. Hence show that the transition rate for (210) $\rightarrow$ (100) is $6{\cdot}27\times10^8$ per excited atom per second. (Assemble the necessary information from p. 100.)

3. Show that the electric dipole transition rate for (200) $\rightarrow$ (100) is zero. (The excited state (200) is **metastable** and the transition is **forbidden**: see below on *selection rules*. Actually, the transition rate is about $10^4\,\mathrm{s}^{-1}$, on account of other radiative mechanisms.)

4. The result of problem 2 does not imply that an excited atom radiates $6{\cdot}27\times10^8$ photons every second! It means that the probability of emission of *one* photon in a very short time δt is $\delta t/\tau$, where $\tau^{-1}=6{\cdot}27\times10^8\,\mathrm{s}^{-1}$.

   Between t and $t+\delta t$ the change in the number N of a large collection of excited hydrogen atoms, on account of loss to the lower level, is

$$\delta N = -N\delta t/\tau.$$

   Show that this implies $N=N_0\exp(-t/\tau)$, and therefore that τ is an estimate of the *lifetime* of the excited state. (The lifetimes of atomic levels are typically in the range 10^{-9} to 10^{-7} s in the optical region, but may be very much longer in the infrared and microwave regions, on account of the factor ω_0^3 in the formula for the transition rate.)

Natural line-width

The frequency of the spectral line for the transition (210) $\rightarrow$ (100) in atomic hydrogen is $\omega_0/2\pi=2{\cdot}46\times10^{15}$ Hz, while the lifetime $\tau=1{\cdot}59\times10^{-9}$ s (see problems above). Now, the longer the lifetime, the more accurately the frequency can be established. In this case, about four million oscillations at frequency $\omega_0/2\pi$ can be fitted into one lifetime τ; we therefore expect that the best experimental observations of this spectral line will show an irreducible **natural line-width** of something like one part in a few million, not untypical of lines in the ultraviolet region. This incidentally shows what a poor radiator an atom is; from the historical point of view, this is just as well, since a more efficient radiator would have emitted *more diffuse* lines, and the very precise spectroscopic measurements which were so essential for the early development of quantum mechanics might not have been forthcoming.

By the way, the hyperfine splitting in the hydrogen atom is also of the order of one part in a million, and is therefore just large enough not to be masked by the natural width of each component in the split line.

PROBLEM

Show that, for the (210) → (100) transition of the hydrogen atom, the emitted wavelength is about 2300 a_0. Comment on the efficiency of the atom as an 'antenna'.

Selection Rules

Many atoms and molecules are highly symmetric in a geometrical sense. The hydrogen atom has a spherically symmetric potential, methane (CH_4) shows a tetrahedral symmetry, and so on. It may well happen as a result that the overwhelming majority of dipole matrix elements turn out to be identically zero, and therefore that the corresponding spectral lines are absent—or are at best only faintly present by way of one of the weaker mechanisms. In such circumstances, for a non-zero electric dipole matrix element the associated transition must meet certain conditions: the **selection rules** for the system. Transitions which fail to satisfy the conditions are **forbidden**; this does *not* mean that they cannot occur, but rather that when they do occur they proceed by way of one of the other, weaker, mechanisms.

One of the most important instances of selection rules occurs for a particle in a spherically symmetric potential. The stationary bound states are indexed by the usual quantum numbers n, l, m, so that for the transition the relevant dipole matrix element is

$$\begin{aligned} \boldsymbol{d} &= \langle n'l'm'|\mathbf{D}|nlm\rangle \\ &= -e\int r^2 \sin\theta \,\mathrm{d}r\,\mathrm{d}\theta\,\mathrm{d}\phi\,.\,R^*_{n'l'}Y^*_{l'm'}\,.\,\boldsymbol{r}\,.\,R_{nl}Y_{lm} \\ &= -e\int_0^\infty r^3\,\mathrm{d}r R^*_{n'l'}R_{nl}\,.\,\boldsymbol{J}_{l'm',\,lm} \end{aligned}$$

where the angular integration

$$\boldsymbol{J}_{l'm',\,lm} = \iint \sin\theta\,\mathrm{d}\theta\,\mathrm{d}\phi\,Y^*_{l'm'}(\theta,\phi)Y_{lm}(\theta\phi)(\sin\theta\cos\phi,\sin\theta\sin\phi,\cos\theta)$$

is singled out for special attention. Now Y_{lm} and $Y^*_{l'm'}$ contain factors $e^{im\phi}$ and $e^{-im'\phi}$ respectively, and inspection of the rest of the integrand shows that the ϕ-integration yields zero unless

$$m-m' = 1, 0, \quad \text{or} \quad -1.$$

It is not so easy to see, but it is nevertheless true, that when the ϕ-integration does not yield zero, then the θ-integration does—unless, that is, we also have

$$l - l' = 1 \quad \text{or} \quad -1.$$

(This is a basic property of spherical harmonics whose proof will be found in any appropriate text on mathematical methods.) Thus we see that a consequence of spherical symmetry is that electric dipole emission occurs only for those transitions satisfying the selection rules

$$\begin{aligned} l &\ \text{ changes by } \pm 1, \quad \text{and} \\ m &\ \text{ changes by } 0 \quad \text{or} \quad \pm 1. \end{aligned}$$

From the physical point of view, these requirements should not surprise us. The angular momentum carried by a photon has magnitude $\hbar$, so it is unlikely that the angular momentum of the emitting system should change by more than $\hbar$. Thus we may reasonably expect that l and m should not change by more than 1. The fact that l *must* change will only be elicited by the properties of spherical harmonics.

PROBLEMS

1. A charged particle executes simple harmonic oscillation along the s-axis with angular frequency ω; the electric dipole moment is

$$\mathsf{D} = q\mathsf{s} = \frac{q}{\sqrt{2}}(\mathsf{a}+\mathsf{a}^{+})$$

(p. 151). Show that $\langle m|\mathsf{D}|n\rangle$ is non-zero only if $m = n \pm 1$. Deduce that the emission spectrum is a *single line* of angular frequency ω. (Refer to problem 3, p. 132.)

2. A rigid diatomic molecule with moment of inertia I and a permanent electric dipole moment along its axis is free to rotate about its centre of mass. The energy levels are $E_l = \hbar^2 l(l+1)/2I$. Give reasons for supposing that the selection rule for this system is

$$l - l' = \pm 1.$$

Deduce that the spectrum consists of a band of equally spaced lines, with angular frequencies $\hbar l/I$ ($l \geqslant 1$). (Such molecules exist; HCl is a good example. The rotational spectrum itself is in the far infrared, but the relevant transitions can occur simultaneously with electronic or vibrational transitions, which are split into sets of equally spaced lines as a result.)

3. Verify the selection rules for the spherical harmonics on p. 81–2.

Anisotropic radiation

Radiation from an excited gas is expected to be isotropic and unpolarized even if the emitters themselves are not symmetric, since the *orientations* of the emitters are random. If however the emitters are regularly oriented by some means—embedding in a lattice, or polarization by an inhomogeneous field in the Stern–Gerlach manner—the nature of the radiation will depend on the direction of emission. Even if the emitters themselves are spherically symmetric, a uniform external field will reduce the symmetry of the total system substantially, and the radiation will again be anisotropic. This has evident implications for the Stark and Zeeman effects.

General prescriptions for the intensity and the polarization of the radiation in the direction $\boldsymbol{u}$ are easily obtained. For the *intensity*, we simply refrain from doing the final integration over the solid angle (p. 188):

transition rate for emission in the direction $\boldsymbol{u}$

$$= \frac{\omega_0^3}{8\pi^2\varepsilon_0\hbar c^3}\{\boldsymbol{d}\,.\,\boldsymbol{d}^* - (\boldsymbol{d}\,.\,\boldsymbol{u})(\boldsymbol{d}^*\,.\,\boldsymbol{u})\} \text{ photons per excited atom per second per unit solid angle}$$

As for the *polarizations*, we rewrite this transition rate as

$$\frac{\omega_0^3}{8\pi^2\varepsilon_0\hbar c^3}\{|\boldsymbol{d}\,.\,\boldsymbol{\lambda}_1^*|^2 + |\boldsymbol{d}\,.\,\boldsymbol{\lambda}_2^*|^2\}$$

(p. 187), and interpret the two terms as probabilities. Take as the two basic polarization vectors

$$\boldsymbol{\lambda}_1 \quad \text{parallel to } \boldsymbol{d} - \boldsymbol{u}(\boldsymbol{d}\,.\,\boldsymbol{u}),$$

$$\boldsymbol{\lambda}_2 \quad \text{parallel to } \boldsymbol{u} \wedge \boldsymbol{d}^*,$$

a legitimate choice as $\boldsymbol{\lambda}_1, \boldsymbol{\lambda}_2, \boldsymbol{u}$ are orthogonal in pairs. Now imagine a polarizer which *certainly* transmits a photon in polarization state $\boldsymbol{\lambda}_2$; the probability that the radiation from $\boldsymbol{d}$ gets through the polarizer is proportional to $|\boldsymbol{d}\,.\,\boldsymbol{\lambda}_2^*|^2$, which is *zero* since $\boldsymbol{d}\,.\,(\boldsymbol{u} \wedge \boldsymbol{d}) = 0$. The radiation is therefore in the orthogonal polarization state, represented by $\boldsymbol{\lambda}_1$.

Thus both the intensity and the polarization are determined by the component of $\boldsymbol{d}$ perpendicular to $\boldsymbol{u}$, a result also true for a *classical* radiating dipole.

PROBLEMS

1. Show that
   (a) the component of $\boldsymbol{d}$ perpendicular to $\boldsymbol{u}$ is $\boldsymbol{d} - \boldsymbol{u}(\boldsymbol{d}\,.\,\boldsymbol{u})$;
   (b) $\boldsymbol{\lambda}_1\,.\,\boldsymbol{u} = \boldsymbol{\lambda}_2\,.\,\boldsymbol{u} = 0$; $\boldsymbol{\lambda}_2^*\,.\,\boldsymbol{\lambda}_1 = 0$;
   (c) $|\boldsymbol{d} - \boldsymbol{u}(\boldsymbol{d}\,.\,\boldsymbol{u})|^2 = \boldsymbol{d}\,.\,\boldsymbol{d}^* - (\boldsymbol{d}\,.\,\boldsymbol{u})(\boldsymbol{d}^*\,.\,\boldsymbol{u})$.

2. Show that one may write

$$\boldsymbol{d}-\boldsymbol{u}(\boldsymbol{d}\,.\,\boldsymbol{u}) = A_{\boldsymbol{u}}^{\frac{1}{2}}\boldsymbol{\lambda}_{\boldsymbol{u}}$$

where the intensity per steradian per atom and polarization for the direction $\boldsymbol{u}$ are

$$\omega_0^4 A_{\boldsymbol{u}}/8\pi^2\varepsilon_0 c^3 \quad \text{and} \quad \boldsymbol{\lambda}_{\boldsymbol{u}}.$$

The Stark and Zeeman effects in the hydrogen atom

A hydrogen atom in a uniform electric or magnetic field has lost its spherical symmetry. Consequently, certain degeneracies are lifted (pp. 104–5). A further consequence is that the components of the resultant spectral splitting have intensities and polarizations which depend on the direction of emission. We obtain the details through a straightforward application of the formula of the last section; we shall do this for the $n = 2 \rightarrow n = 1$ transitions.

As a preliminary, it is useful to list the relevant dipole matrix elements:

$$\langle 100|\mathbf{D}|211\rangle = -0{\cdot}745\, ea_0 2^{-\frac{1}{2}}(1, \mathrm{i}, 0)$$

$$\langle 100|\mathbf{D}|210\rangle = -0{\cdot}745\, ea_0 (0, 0, 1) \qquad \text{(cf. problem 2, p. 189)}$$

$$\langle 100|\mathbf{D}|21-1\rangle = -0{\cdot}745\, ea_0 2^{-\frac{1}{2}}(-1, \mathrm{i}, 0)$$

$$\langle 100|\mathbf{D}|200\rangle = 0 \text{ (by the selection rule } \Delta l = \pm 1).$$

In the rest of this section, we shall drop the common factor $-0{\cdot}745\, ea_0$ in these expressions and also the factor $\omega_0^3/8\pi^2\varepsilon_0\hbar c^3$ in the transition rate, since this hardly changes from component to component in the split line. We shall write $\sim$ to indicate equality after these factors are taken into account; thus,

$$\langle 100|\mathbf{D}|211\rangle \sim 2^{-\frac{1}{2}}(1, \mathrm{i}, 0),$$

and so on. Everything depends on the unit or zero vectors on the right.

The Zeeman effect (p. 165) is the simpler of the two. For a field in the z-direction, there are three possible transitions (actually four: one is forbidden). For the transition $(211) \rightarrow (100)$ the relevant dipole matrix element is $\boldsymbol{d} \sim 2^{-\frac{1}{2}}\,(1, \mathrm{i}, 0)$. If we view the resulting spectral line along the x-direction, then $\boldsymbol{u} = (1, 0, 0)$ and

$$A^{\frac{1}{2}}\boldsymbol{\lambda} = \boldsymbol{d}-\boldsymbol{u}(\boldsymbol{d}\,.\,\boldsymbol{u}) \sim 2^{-\frac{1}{2}}\,(0, \mathrm{i}, 0).$$

Hence (see problem 2 above),

$$\boldsymbol{\lambda} = (0, \mathrm{i}, 0), \quad \text{and} \quad A \sim \tfrac{1}{2}$$

interpreted immediately as a plane-polarization along the y-axis, and a relative intensity of $\frac{1}{2}$. Viewing along the z-direction implies $\boldsymbol{u} = (0, 0, 1)$,

$$A^{\frac{1}{2}}\boldsymbol{\lambda} \sim 2^{-\frac{1}{2}}\,(1, \mathrm{i}, 0),$$

and

$$\lambda = 2^{-\frac{1}{2}}(1, \mathrm{i}, 0), \quad A \sim 1;$$

that is, the *same* component viewed along the z-direction is now *circularly* polarized, and is now twice as strong. This information is summarized in the first row of the table below; the other rows are established in the same way, while the diagrams give the same information in a more pictorial form.

		Viewing parallel to $\boldsymbol{B}$ $\boldsymbol{u}=(0,0,1)$		Viewing perpendicular to $\boldsymbol{B}$ $\boldsymbol{u}=(1,0,0)$	
Transition	$\boldsymbol{d}\sim$	$\boldsymbol{\lambda}$	$A\sim$	$\boldsymbol{\lambda}$	$A\sim$
(211) → (100)	$(1,\mathrm{i},0)/2^{\frac{1}{2}}$	$(1,\mathrm{i},0)/2^{\frac{1}{2}}$	1	$(0,\mathrm{i},0)$	$\frac{1}{2}$
(210) → (100)	$(0,0,1)$	–	0	$(0,0,1)$	1
(200) → (100)	0 (forbidden)	–	0	–	0
(21-1) → (100)	$(-1,\mathrm{i},0)/2^{\frac{1}{2}}$	$(-1,\mathrm{i},0)/2^{\frac{1}{2}}$	1	$(0,\mathrm{i},0)$	$\frac{1}{2}$

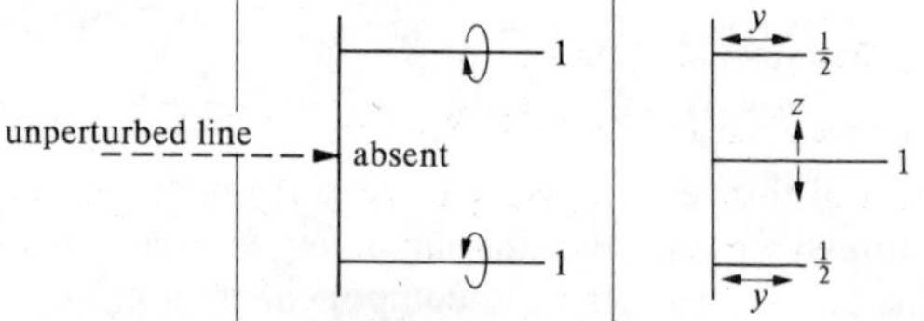

The normal Zeeman effect for hydrogen atom transitions from $n=2$ to $n=1$ and $\boldsymbol{B}=(0,0,B)$

The Stark effect (p. 163) is complicated somewhat by a residual degeneracy, which leaves the transitions $(211) \rightarrow (100)$ and $(21-1) \rightarrow (100)$ unseparated. Viewed from the z-direction, these transitions yield circular polarization in two opposite senses. A gas of excited hydrogen atoms will undergo these transitions in equal numbers; consequently, this spectral line is completely unpolarized. In addition, the other two lines are absent. Thus, viewed from the z-direction, the Stark effect cannot be seen. The details for the two viewing directions are given in the table on p. 195, using the excited levels already obtained by perturbation theory.

We have ignored electron spin throughout. However, the electron intrinsic magnetic moment is irrelevant to the electric dipole moment of the atom, and therefore has no effect at the level we have been considering. Of course, the whole picture is changed when we come to consider fine structure (p. 106) or hyperfine structure (p. 169), where electron spin and magnetic moment play a crucial part.

		Viewing parallel to $\mathscr{E}$ $\boldsymbol{u}=(0,0,1)$		Viewing perpendicular to $\mathscr{E}$ $\boldsymbol{u}=(1,0,0)$	
Transition	$\boldsymbol{d}\sim$	$\boldsymbol{\lambda}$	$A\sim$	$\boldsymbol{\lambda}$	$A\sim$
$2^{-\frac{1}{2}}((200)+(210)) \to (100)$	$(0,0,1)/2^{\frac{1}{2}}$	—	0	$(0,0,1)$	$\frac{1}{2}$
degenerate $(211) \to (100)$	$(1,i,0)/2^{\frac{1}{2}}$	$(1,i,0)/2^{\frac{1}{2}}$	1	$(0,i,0)$	$\frac{1}{2}$
degenerate $(21-1) \to (100)$	$(-1,i,0)/2^{\frac{1}{2}}$	$(-1,i,0)/2^{\frac{1}{2}}$	1	$(0,i,0)$	$\frac{1}{2}$
$2^{-\frac{1}{2}}((200)-(210)) \to (100)$	$(0,0,1)/2^{\frac{1}{2}}$	—	0	$(0,0,1)$	$\frac{1}{2}$

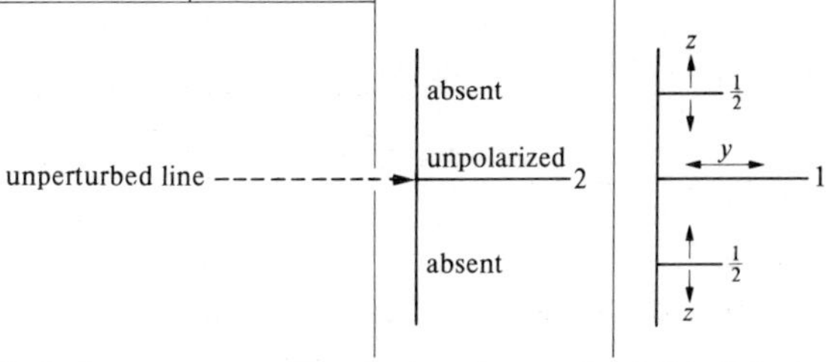

The Stark effect for hydrogen atom transitions from $n = 2$ to $n = 1$ and $\mathscr{E} = (0, 0, \mathscr{E})$

PROBLEMS

1. Verify the values of $\langle 100|\mathbf{D}|211\rangle$, etc.
2. Discuss the Zeeman and Stark effects for other viewing directions.
3. A hydrogen atom is simultaneously perturbed by parallel fields $\mathscr{E} = (0, 0, \mathscr{E})$ and $\boldsymbol{B} = (0, 0, B)$. Obtain the diagrams:

Viewing parallel to $\mathscr{E}$ and $\boldsymbol{B}$ Viewing perpendicular to $\mathscr{E}$ and $\boldsymbol{B}$.

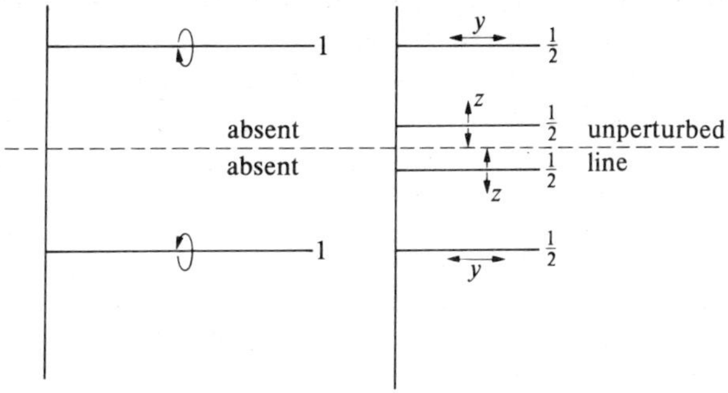

(Refer to problem 1, p. 166.)

4. Show that, when the z-component σ_z of electron spin is taken as a fourth quantum number,

$$\langle nlm\sigma_z|\mathbf{D}|n'l'm'\sigma_z'\rangle = \langle nlm|\mathbf{D}|n'l'm'\rangle\langle\sigma_z|\sigma_z'\rangle$$

(compare p. 125), and hence that the selection rule for σ_z is $\Delta\sigma_z = 0$. (It follows that even though the electron magnetic moment has a substantial potential energy in an external magnetic field, electric dipole transitions will never show this up. Thus the Zeeman effect is *normal*, not anomalous, in this case.)

Stimulated transitions

So far we have been dealing with the behaviour of an excited system in the dark. Let us now turn to the other extreme: a system in a strong oscillating external electromagnetic field. The discussion on resonance (p. 179) shows that the external field will have little effect on the system unless its frequency ω (or rather $\hbar\omega$) matches the gap between two energy levels. Let us suppose there is *one* such match, $E_1 - E_2 = \hbar\omega$, and use the resonance approximation.

The perturbing term in the Hamiltonian is

$$\mathsf{H}' = -\mathbf{D}\,.\,\mathscr{E}(0) = -\mathbf{D}\,.\,(\boldsymbol{Z}\exp(-\mathrm{i}\omega t) + \boldsymbol{Z}^*\exp(\mathrm{i}\omega t))$$

(cf. p. 147). The two-level resonance approximation (p. 180) may immediately be applied, and requires in this case that

$$W = -\langle E_1|\mathbf{D}|E_2\rangle\,.\,\boldsymbol{Z} = -\boldsymbol{d}^*\,.\,\boldsymbol{Z}, \text{ say,}$$

and the effective Hamiltonian for the transition is

$$\begin{pmatrix} 0 & -\boldsymbol{d}^*\,.\,\boldsymbol{Z} \\ -\boldsymbol{d}\,.\,\boldsymbol{Z}^* & 0 \end{pmatrix}.$$

The effect of the external radiation is to make the state of the system swing to and fro between the levels, the time to switch between states being $\pi\hbar/2|\boldsymbol{d}^*\,.\,\boldsymbol{Z}|$. This time is generally very much shorter than the lifetime for spontaneous emission. The formula takes into account not only the magnitudes of the dipole moment $\boldsymbol{d}$ and the external field $\boldsymbol{Z}$, but also any polarization effects there may be, when these are important.

The switch from the lower to the upper level requires a supply of energy $\hbar\omega$ from the external field: this is referred to as **absorption**. The opposite transition is **stimulated emission**, where the external field gains energy from the system. The fact that the amount of energy transferred is $\hbar\omega$ in each case suggests that it is a *single photon* that is being absorbed or emitted. This can be verified by a more complete treatment in which the electromagnetic field is dealt with as a full quantum system. The relevant *quantum field theory* is outside of the range of this book, however.

Coherence

When an atom or molecule undergoes stimulated emission, the emitted radiation is *in phase* with the stimulating field. (If this were not so, some destructive interference would occur, the amplitude of the stimulating field would not increase as much as it should, and an energy imbalance would then have to be explained.) This is **temporal coherence**, and is crucial to laser action.

In principle, a laser is a collection of atoms or other similar systems with a **population inversion**: a *majority* of the atoms are in an excited level E_1. Any stray radiation of the correct frequency will stimulate coherent emission from the excited atoms, and will be amplified as a result. This process will continue until the E_1 and E_2 populations are equal; it will then stop, as the competing process of absorption will then be going at the same rate. On account of the temporal coherence, the amplified radiation is extraordinarily regular, its bandwidth being extremely narrow.

If a means can be found to maintain the population inversion in some other way ('**pumping**'), the amplifier becomes an *oscillator* provided precautions are taken to maintain stability. It is these devices which are usually called 'lasers' (the name is itself an acronym: **l**ight **a**mplification by **s**timulated **e**mission of **r**adiation). Their forerunners were the *masers*, which operated at the very much lower **m**icrowave frequencies, by exactly the same kind of mechanism.

The frequency stability of such oscillators (particularly those using microwave transitions) is so impressively good that, as **atomic clocks**, they now provide our primary frequency and timekeeping standards. Currently, their stability may be of the order of a few milliseconds per millenium.

16. Variational approximations

The underlying principle

PERTURBATION methods have their limitations. As we have seen, it is certainly true that there are many physical situations which may be assessed with high accuracy simply because some parameter or other is *small*. It is equally true that there are other situations where there is no such small parameter. Perturbation approximations may then fail completely, and alternative approaches must be found.

We shall consider just one such approach—the **variational method**—and shall apply it to just one type of problem, the estimation of the ground state energy eigenvalue of a physical system.

The basic principle is simple. Suppose that E_0 is the *smallest* eigenvalue of the Hamiltonian H. Then, for any state $|\psi\rangle$ whatever, the mean value for the energy satisfies

$$\langle \mathsf{H} \rangle_\psi \geqslant E_0;$$

this is an everyday statistical requirement, quite apart from quantum mechanics. It follows that

$$\frac{\langle \psi | \mathsf{H} | \psi \rangle}{\langle \psi | \psi \rangle} \geqslant E_0,$$

the left side now being the quantum mechanical definition of mean value. (Note that, for reasons of convenience, we are allowing the state $|\psi\rangle$ to be unnormalized.) This simple inequality is the basis of all the developments of this chapter. Note that equality is obtained only if $|\psi\rangle$ is the ground state.

Now it may happen that, even if the problem is too complex for us to find the ground state exactly, we may nevertheless be able to give good reasons why some 'simpler' state $|\psi\rangle$ may be supposed to be a close approximation to the ground state. In this case, we may use

$$E_{\text{est}} = \frac{\langle \psi | \mathsf{H} | \psi \rangle}{\langle \psi | \psi \rangle}$$

as an *estimate* for the ground state energy; because of the basic inequality, it will always be an *overestimate*.

In practice, we proceed as follows. First, we need to use any physical insight we may have to assess the kind of shape that the ground state wavefunction will have. Then we *choose* a function with two requirements:

(i) it must have the right kind of shape to be a reasonable approximation to the ground state; and
(ii) it must be simple enough for us to handle.

More generally, we may choose a *family* of functions $\psi(x;\alpha)$, indexed by one or more continuous parameters α, and evaluate $E_{\text{est}}(\alpha)$ for each. Finally, we **vary** the parameter α to find the absolute minimum of $E_{\text{est}}(\alpha)$; this will be our estimate for E_0.

At first sight it may seem odd that such a crude approach should work at all well. However, if our choice of the **trial function** $\psi(x;\alpha)$ is reasonably near the ground state function, then E_{est} may be *very* near E_0. The reason is that E_0 is the absolute minimum of $\langle H\rangle$, and in the vicinity of its minimum, $\langle H\rangle$ is insensitive to small errors in the ground state wavefunction.

The requirement that our choice must be simple enough to handle is important, especially if the calculations are to be done 'by hand'. Variational methods are often tedious and long-winded, and may require dedication of the highest order! Typically, many integrals may have to be evaluated, and the minimization process itself may be very involved. On the other hand, matters can be arranged so that the work is very repetitive in character, permitting the use of an electronic computer. Colossal amounts of computer time have been consumed in the last 30 years in the search for ever more accurate ground state energies of atoms and molecules.

So that the wood should not be obscured by the trees, much of the algebraic work will be taken for granted. Obviously, however, the author has had to do it in full; it will do the reader good to follow his example!

The particle in a box

Let us begin with a system whose solution we know (p. 61). A particle is confined in one dimension to the region $0 \leqslant x \leqslant 1$. The requirements on the energy eigenfunctions $\psi(x)$ are

$$-\frac{d^2\psi}{dx^2} = E\psi, \quad \psi(0) = 0, \quad \psi(1) = 0.$$

(We have dropped all the physical constants: they make no difference to the discussion.) We know the solution for the ground state,

$$E_0 = \pi^2, \quad \psi_0(x) \sim \sin \pi x.$$

What if this solution were unknown to us? Then we would use our physical 'feel' for the problem: ψ_0 must be zero at the ends of the range, and we remember that the ground state is the one which oscillates least as x

varies. We may therefore wish to try something like

$$\psi = x(1-x),$$

which has the right kind of shape. In fact, we shall widen the possibilities considerably by adopting as the trial function (unnormalized)

$$\langle x|\psi_\alpha\rangle = \psi(x;\alpha) = x^\alpha(1-x)^\alpha. \qquad (\alpha > 0)$$

We shall vary the parameter α to obtain the best estimate for E_0.

The calculation goes as follows. First,

$$\mathsf{H}|\psi_\alpha\rangle \leftarrow -\frac{d^2}{dx^2}\{x^\alpha(1-x)^\alpha\}$$

$$= -\alpha(\alpha-1)x^{\alpha-2}(1-x)^\alpha + 2\alpha^2 x^{\alpha-1}(1-x)^{\alpha-1} - \alpha(\alpha-1)x^\alpha(1-x)^{\alpha-2}$$

Hence

$$\langle\psi_\alpha|\mathsf{H}|\psi_\alpha\rangle = \int_0^1 dx\, x^\alpha(1-x)^\alpha\{-\alpha(\alpha-1)x^{\alpha-2}(1-x)^\alpha + 2\alpha^2 x^{\alpha-1}(1-x)^{\alpha-1} - \alpha(\alpha-1)x^\alpha(1-x)^{\alpha-2}\}.$$

Also,

$$\langle\psi_\alpha|\psi_\alpha\rangle = \int_0^1 dx\, x^{2\alpha}(1-x)^{2\alpha}.$$

It is convenient to introduce the notation for the integral

$$\mathrm{I}(p,q) = \int_0^1 dx\, x^p(1-x)^q,$$

so that

$$\langle\psi_\alpha|\mathsf{H}|\psi_\alpha\rangle = -\alpha(\alpha-1)\mathrm{I}(2\alpha-2, 2\alpha) + 2\alpha^2\mathrm{I}(2\alpha-1, 2\alpha-1) - \alpha(\alpha-1)\mathrm{I}(2\alpha, 2\alpha-2),$$

$$\langle\psi_\alpha|\psi_\alpha\rangle = \mathrm{I}(2\alpha, 2\alpha).$$

Fortunately, there exist useful relations between 'adjacent' I-integrals,

$$q\mathrm{I}(p, q-1) = (p+q+1)\mathrm{I}(p,q) = p\mathrm{I}(p-1, q);$$

these enable us to reduce the mean energy for the state $|\psi_\alpha\rangle$ to the simple form

$$\langle H\rangle_\alpha = \frac{\langle\psi_\alpha|\mathsf{H}|\psi_\alpha\rangle}{\langle\psi_\alpha|\psi_\alpha\rangle} = \frac{2\alpha(4\alpha+1)}{2\alpha-1}.$$

Any value of $\alpha > \frac{1}{2}$ yields a value of $\langle H \rangle_\alpha$ which we know must be an overestimate of E_0. In the spirit of the method, we must choose α to give the lowest such overestimate. The condition that $\langle H \rangle_\alpha$ should be stationary is

$$\frac{\mathrm{d}\langle H \rangle_\alpha}{\mathrm{d}\alpha} = 0;$$

this reduces to

$$8\alpha^2 - 8\alpha - 1 = 0,$$

whose positive root $\alpha = 1{\cdot}112\,37\ldots$ gives the minimum

$$\langle H \rangle_\alpha = 9{\cdot}90.$$

The actual ground state energy is $E_0 = \pi^2 = 9{\cdot}87$.

The fact that the variational estimate is in error by only $\frac{1}{3}$ per cent follows from a fortunate accident that the functions

$$\sin^2 \pi x \quad \text{and} \quad x^{2\alpha}(1-x)^{2\alpha} \quad (\alpha = 1{\cdot}11\ldots)$$

are (after normalization) remarkably similar (see figure, where the solid curve is $\sin^2 \pi x$); the maximum discrepancy is about $2\frac{1}{2}$ per cent. Such good agreement for so little effort is unusual!

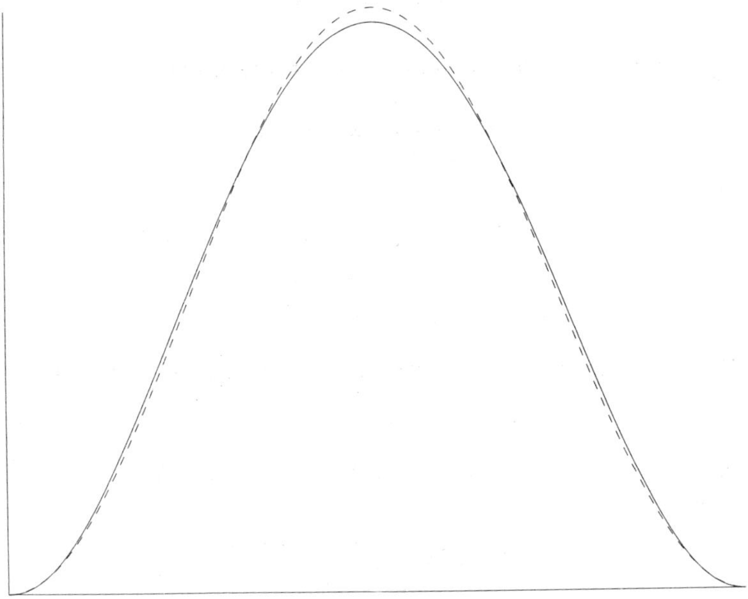

Trial functions with several parameters

It should be clear that the chance of getting a good fit to the true ground-state wavefunction is increased by widening the possibilities for the trial function. One way to do this is to introduce a trial function depending on two (or more) parameters, $\psi(x;\alpha,\beta)$ say, and varying α and β simultaneously to achieve the absolute minimum of $\langle H\rangle_{\alpha\beta}$. Necessary conditions to be satisfied are

$$\frac{\partial\langle H\rangle_{\alpha\beta}}{\partial\alpha}=0,\quad \frac{\partial\langle H\rangle_{\alpha\beta}}{\partial\beta}=0,$$

whose simultaneous solution gives the best values of α and β, and hence the lowest $\langle H\rangle$.

This seems straightforward enough, but can be troublesome in practice. As an example, let us now apply a *uniform force* to the particle in the box, with potential gx. The requirements on ψ are now

$$-\frac{\mathrm{d}^2\psi}{\mathrm{d}x^2}+gx\psi=E\psi,\quad \psi(0)=0,\quad \psi(1)=0.$$

Our feel for the physics of the problem suggests that the ground state wavefunction will no longer be symmetric about $x=\frac{1}{2}$, but will be lopsided in the direction of the applied force. Let us therefore take as trial function

$$\psi(x;\alpha,\beta)=x^{\alpha}(1-x)^{\beta}\quad\text{(unnormalized)},$$

where the disposable parameters α, β are now expected to be different.

To begin with, the work is only slightly more complicated. First,

$$\begin{aligned}\mathsf{H}|\psi_{\alpha\beta}\rangle\leftarrow&-\frac{\mathrm{d}^2}{\mathrm{d}x^2}\{x^{\alpha}(1-x)^{\beta}\}+gx^{\alpha+1}(1-x)^{\beta}\\ =&-\alpha(\alpha-1)x^{\alpha-2}(1-x)^{\beta}+2\alpha\beta x^{\alpha-1}(1-x)^{\beta-1}-\beta(\beta-1)x^{\alpha}(1-x)^{\beta-2}\\ &+gx^{\alpha+1}(1-x)^{\beta},\end{aligned}$$

which leads to

$$\begin{aligned}\langle\psi_{\alpha\beta}|\mathsf{H}|\psi_{\alpha\beta}\rangle=&-\alpha(\alpha-1)\mathrm{I}(2\alpha-2,2\beta)+2\alpha\beta\mathrm{I}(2\alpha-1,2\beta-1)\\ &-\beta(\beta-1)\mathrm{I}(2\alpha,2\beta-2)+g\mathrm{I}(2\alpha+1,2\beta).\end{aligned}$$

Also

$$\langle\psi_{\alpha\beta}|\psi_{\alpha\beta}\rangle=\mathrm{I}(2\alpha,2\beta).$$

Once again it is possible to use the relations between adjacent I-integrals to find

$$\langle H\rangle_{\alpha\beta} = \frac{(\alpha+\beta)(\alpha+\beta-1)(2\alpha+2\beta+1)}{(2\alpha-1)(2\beta-1)} + g\frac{2\alpha+1}{2(\alpha+\beta+1)},$$

after relatively few lines of work.

It is when we embark on the process of minimization that matters begin to get out of hand. To locate the minimum of even the fairly simple function we have here is troublesome by algebraic methods, and it is usual to use more or less *ad hoc* computer methods for trial functions with more than two or three disposable parameters. A convenient method where there are just *two* parameters is to use a standard computer contouring program to draw the curves of constant $\langle H\rangle_{\alpha\beta}$ ('contours') in the $\alpha\beta$-plane. Two such diagrams are shown in the figure for the present problem with $g = 0$ and $g = 10$; the location of the minimum is clearly visible. Of course, pictorial methods of this kind fail when there are more than a very few parameters.

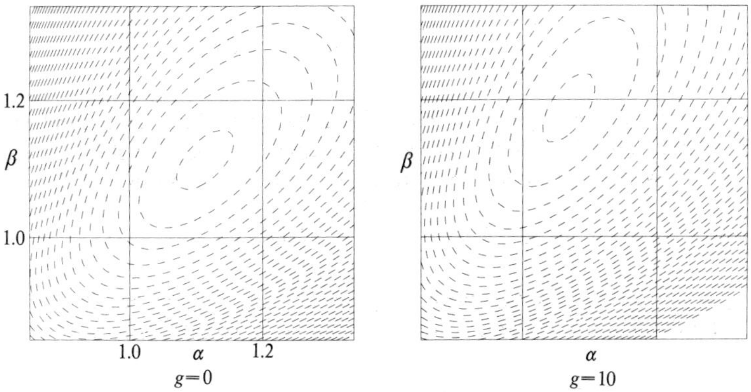

As a general rule, it is true to say that trial functions with many parameters cause trouble, and they are hardly ever used. There is, however, one crucially important exception, which forms the subject of the next section.

PROBLEMS

1. Verify that

$$\frac{\mathrm{d}}{\mathrm{d}x}\{x^p(1-x)^q\} = px^{p-1}(1-x)^{q-1} - (p+q)x^p(1-x)^{q-1}.$$

Deduce the relations between the adjacent I-integrals.

2. Fill in the algebraic details of the last two sections.

3. Carry out—as far as you are prepared—the minimization of $\langle H\rangle_{\alpha\beta}$ by any pencil-and-paper method you can think of.

Linear combinations as trial functions

In applying the variational method to a Hamiltonian H, one possible choice of trial state is a *linear combination*

$$|\psi\rangle = |1\rangle\alpha_1 + |2\rangle\alpha_2 + \ldots + |N\rangle\alpha_N$$

where the N states $|1\rangle \ldots |N\rangle$ are as usual chosen according to our feel for the problem, and the N parameters $\alpha_1 \ldots \alpha_N$ are to be varied to minimize $\langle H\rangle$. With a trial state of this kind, there is a general recipe, as follows.

Evaluate the N^2 expressions $\langle m|\mathsf{H}|n\rangle = H_{mn}$, and assemble them in the obvious way into a Hermitian matrix $\boldsymbol{H}$. Do the same for the further N^2 expressions $\langle m|n\rangle = J_{mn}$. (Observe, by the way, that we do not require the states $|n\rangle$ to be normalized, or even to be mutually orthogonal; thus $\boldsymbol{J}$ need not be the unit matrix, or even diagonal.) Then

$$\langle\psi|\mathsf{H}|\psi\rangle = \sum_{m=1}^{N}\sum_{n=1}^{N}(\alpha_m^*\langle m|)\mathsf{H}(|n\rangle\alpha_n^*) = \sum_{mn}\alpha_m^* H_{mn}\alpha_n;$$

similarly

$$\langle\psi|\psi\rangle = \sum_{mn}\alpha_m^* J_{mn}\alpha_n.$$

In performing the minimization, it is best to introduce a *Lagrange multiplier* E: the condition that $\langle H\rangle$ is a minimum is equivalent to the condition that $\langle\psi|\mathsf{H}|\psi\rangle$ be a minimum, subject to $\langle\psi|\psi\rangle = 1$. The Lagrange method allows us to take as equivalent the condition that

$$F = \langle\psi|\mathsf{H}|\psi\rangle - E\langle\psi|\psi\rangle$$

be unconditionally minimized. For a trial function of the kind considered here

$$F = \sum_{mn}\alpha_m^*(H_{mn} - EJ_{mn})\alpha_n,$$

the conditions for a minimum therefore being

$$\frac{\partial F}{\partial\alpha_m^*} = \sum_{n}(H_{mn} - EJ_{mn})\alpha_n = 0 \text{ for each } m = 1 \ldots N.$$

This set of N homogeneous linear equations in $\alpha_1 \ldots \alpha_N$ can have a non-trivial solution only if the **secular determinant** is zero:

$$\det|\boldsymbol{H} - E\boldsymbol{J}| = 0.$$

As is usual whenever a Lagrange multiplier is introduced in theoretical physics, E has an important significance. By inspection, whenever the αs satisfy the requirements, F is *zero*. Thus

$$\langle\psi|\mathsf{H}|\psi\rangle = E\langle\psi|\psi\rangle,$$

that is, E is in fact the estimate we seek. This completes the **recipe** for this kind of variational calculation; to summarize:

(i) Choose a suitable linear combination

$$|\psi\rangle = \sum_n |n\rangle\alpha_n$$

as trial state;

(ii) assemble the matrices $\boldsymbol{H}$, $\boldsymbol{J}$, with elements

$$H_{mn} = \langle m|\mathsf{H}|n\rangle,\ J_{mn} = \langle m|n\rangle;\ \text{and}$$

(iii) obtain the lowest root of $\det|\boldsymbol{H}-E\boldsymbol{J}| = 0$. This is E_{est}, the required estimate of ground state energy. (If required, the corresponding values of $\alpha_1 \ldots \alpha_N$ form an eigenvector of the matrix:

$$(\boldsymbol{H} - E_{\text{est}}\boldsymbol{J})\boldsymbol{\alpha} = 0,$$

though this is not often used.)

This is the only situation that I know of when the parameters α can be completely eliminated in any general way. That in itself would be enough to make linear combinations important as trial functions. The outcome is even better than this: after the elimination of the αs, we are left with a condition which gives us E_{est} directly as the root of a determinant. There are well-defined and precise numerical techniques for handling such expressions, even when the order of the determinant is very large. Such considerations have led to a large department of 'theoretical quantum chemistry', in which attempts are made to describe the ground states (and low excited states too) of atoms and molecules as **L**inear **C**ombinations of suitably chosen **A**tomic **O**rbitals (the **LCAO method**). An **orbital** is a supposed possible wavefunction of one of the electrons in the system.

The particle in a box, for the last time

In the light of the last section, let us take a final look at the particle in a box, with the potential gx (p. 202). This time we shall achieve the required lopsidedness by including a linear factor:

$$\psi(x;\alpha_1,\alpha_2) = (\alpha_1 + \alpha_2 x)x^\alpha(1-x)^\alpha.$$

(In what follows, α is no longer a variational parameter, but is kept fixed at $\alpha = 1{\cdot}112\ldots$, the best value for $g = 0$.) In other words, ψ is to be a linear combination of the two 'orbitals'

$$\psi_1(x) = x^\alpha(1-x)^\alpha \quad \text{and} \quad \psi_2(x) = x^{\alpha+1}(1-x)^\alpha.$$

The recipe requires us to evaluate

$$H_{mn} = \int_0^1 \mathrm{d}x\psi_m(x)\left(-\frac{\mathrm{d}^2}{\mathrm{d}x^2}+gx\right)\psi_n(x)$$

and

$$J_{mn} = \int_0^1 \mathrm{d}x\psi_m(x)\psi_n(x),$$

for $m, n = 1, 2$ in all four combinations. In fact, some of these have been evaluated already (cf. p. 202); the others are dealt with similarly. When all eight integrals have been expressed as multiples of $\mathrm{I}(2\alpha, 2\alpha)$ (p. 200), we find

$$\boldsymbol{H} = \begin{pmatrix} \dfrac{2\alpha(4\alpha+1)}{2\alpha-1}+g\cdot\frac{1}{2} & \dfrac{\alpha(4\alpha+1)}{2\alpha-1}+g\,\dfrac{\alpha+1}{4\alpha+3} \\ \dfrac{\alpha(4\alpha+1)}{2\alpha-1}+g\cdot\dfrac{\alpha+1}{4\alpha+3} & \dfrac{2\alpha(\alpha+1)}{2\alpha-1}+g\,\dfrac{2\alpha+3}{4(4\alpha+3)} \end{pmatrix}\mathrm{I}(2\alpha, 2\alpha),$$

$$\boldsymbol{J} = \begin{pmatrix} 1 & \frac{1}{2} \\ \frac{1}{2} & \dfrac{\alpha+1}{4\alpha+3} \end{pmatrix}\mathrm{I}(2\alpha, 2\alpha).$$

All that remains is to insert the value $\alpha = 1{\cdot}112\ldots$ throughout, and to solve the equation

$$\det|\boldsymbol{H}-E\boldsymbol{J}| = 0,$$

which in this case reduces to a quadratic for E. The required E_{est} is the lower root.

Incidentally, there is no objection to regarding α also as a parameter to be varied, if we so wish. When $g \neq 0$, the optimum value of α will be different from $1{\cdot}112\ldots$, and can be found by minimizing the E_{est} just found—which obviously now depends on α.

PROBLEMS

1. For the system governed by

$$-\frac{\mathrm{d}^2\psi}{\mathrm{d}x^2}+gx\psi = E_0\psi, \quad \psi(0) = 0, \quad \psi(1) = 0,$$

which of the following trial functions has the right kind of shape? Rank the acceptable ones for ease of application.

$\alpha_1 \sin \pi x + \alpha_2 \sin 2\pi x$ $\qquad$ $\alpha_1 \sin \pi x + \alpha_2 \cos \pi x$

$\sin \alpha_1 x \sin \alpha_2(1-x)$ $\qquad$ $(1+\alpha x) \sin \pi x$

$e^{\alpha x} \sin \pi x.$

2. A particle confined to the positive x-axis experiences a constant force $-mg$. The Schrödinger equation is

$$-\frac{\hbar^2}{2m}\frac{d^2\psi}{dx^2}+mgx\psi = E\psi,$$

with

$$\psi(0)=0,\quad \psi(x)\to 0 \quad \text{as} \quad x\to\infty.$$

Justify the use of $xe^{-\alpha x}$ as a trial function, and show that the best value of α is $(3m^2g/2\hbar^2)^{1/3}$. (See p. 236 for useful formulae. $|\psi|^2$ is the solid line in the figure.)

3. Repeat problem 2 with the trial function $xe^{-\beta x^2}$, showing that the best value of β is $(2g^2m^4/9\pi\hbar^2)^{1/3}$. (See p. 236 for useful formulae. $|\psi|^2$ is the broken line in the figure.)

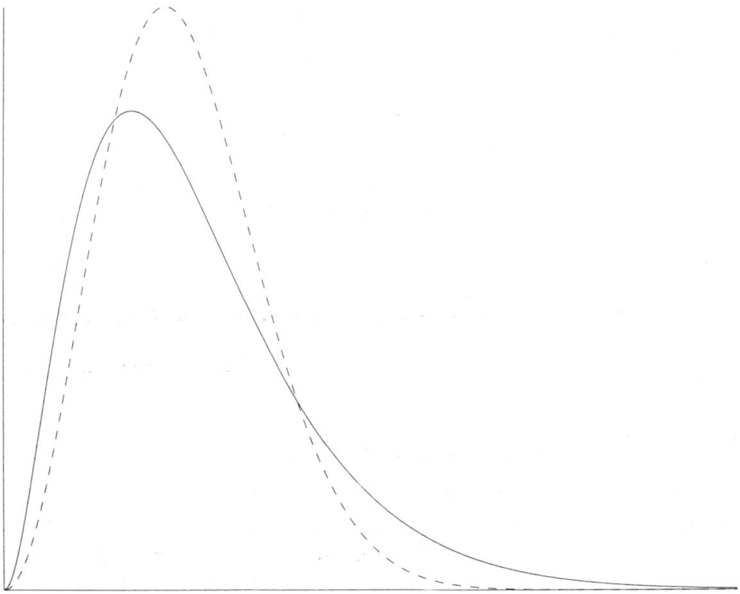

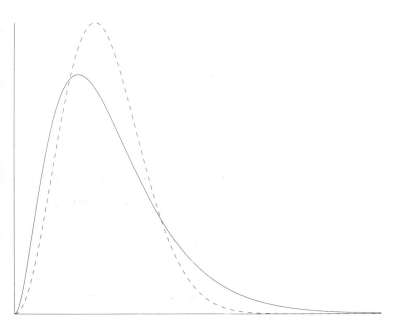

17. Variational approximations: two realistic applications

The ground state of the neutral helium atom

THE neutral helium atom is a good example of how complicated things can get even after a moderate increase in complexity of a physical system. Even if the nucleus of the atom is regarded as fixed at the origin. the two electrons must be treated simultaneously in a quantum-mechanical manner: their joint wavefunction therefore must depend on the position of each, $\psi(\boldsymbol{r}_1, \boldsymbol{r}_2)$. Now this would not be too bad if the electrons were mutually independent, since the Schrödinger equation would then be separable and we would be able to search for special solutions of the kind $\psi(\boldsymbol{r}_1)\psi(\boldsymbol{r}_2)$, for which the two electrons are uncorrelated. But the electrons may be expected to be *strongly* correlated, on account of their mutual repulsion—if one electron is observed to be at a particular position in the atom, it is very unlikely that the second will be observed in the vicinity of the first. Correlations of this kind can be very awkward to handle.

It is convenient to use the scaling introduced in the problem on p. 100, with $Z = 2$. The Hamiltonian for the electrons in the helium atom is

$$\mathsf{H} = \frac{\hbar^2}{2m_e a_0^2}(\mathsf{K}_1 + \mathsf{K}_2 + \mathsf{U})$$

where

$$\mathsf{K}_1 \leftarrow -\nabla_1^2 - \frac{4}{\rho_1}, \quad \mathsf{K}_2 \leftarrow -\nabla_2^2 - \frac{4}{\rho_2}$$

are the contributions from the kinetic energy, and the potential energy in the nuclear Coulomb field, for electrons 1 and 2 respectively, while

$$\mathsf{U} = \frac{2}{|\boldsymbol{\rho}_1 - \boldsymbol{\rho}_2|}$$

is the potential energy of the mutual electrostatic repulsion of the electrons.

Now let us choose a trial function. Consider electron 1 in two extreme situations: when it is very close to the nucleus, it experiences a $Z = 2$ Coulomb force, and when it is a long way off, it experiences the combined effect of the nucleus and the other electron, that is, roughly a $Z = 1$ Coulomb force. Let us therefore adopt as trial function the product of two hydrogen-like wavefunctions

$$\psi_\alpha(\boldsymbol{\rho}_1, \boldsymbol{\rho}_2) = \exp(-\alpha\rho_1)\exp(-\alpha\rho_2),$$

where by the above qualitative argument the 'effective' atomic number α should lie between 1 and 2. In this way, we take some account of the mutual interaction of the two electrons, though they are still uncorrelated.

Let us vary α to obtain the best estimate for the ground-state energy. We need to evaluate several integrals. First,

$$\begin{aligned}\langle\psi_\alpha|\psi_\alpha\rangle &= \int d^3\boldsymbol{\rho}_1 \int d^3\boldsymbol{\rho}_2\, |\exp(-\alpha\rho_1)\exp(-\alpha\rho_2)|^2 \\ &= 16\pi^2 \int_0^\infty \rho_1^2\, d\rho_1 \exp(-2\alpha\rho_1) \,.\, \int_0^\infty \rho_2^2\, d\rho_2 \exp(-2\alpha\rho_2) \\ &= \pi^2\alpha^{-6}.\end{aligned}$$

(Here, and below, we repeatedly use the first formula on p. 236.) Then,

$$\begin{aligned}\mathsf{K}_1|\psi_\alpha\rangle &\leftarrow \left(-\nabla_1^2 - \frac{4}{\rho_1}\right)\exp(-\alpha\rho_1)\exp(-\alpha\rho_2) \\ &= \left(-\alpha^2 + \frac{2\alpha-4}{\rho_1}\right)\exp(-\alpha\rho_1)\exp(-\alpha\rho_2)\end{aligned}$$

(the quickest way to obtain this is to note that

$$\left(-\nabla^2 - \frac{2\alpha}{\rho}\right)\exp(-\alpha\rho) = -\alpha^2\exp(-\alpha\rho);$$

replace α by Z, and see p. 100.) Hence

$$\begin{aligned}\langle\psi_\alpha|\mathsf{K}_1|\psi_\alpha\rangle &= \int d^3\boldsymbol{\rho}_1 \int d^3\boldsymbol{\rho}_2 \exp(-\alpha\rho_1)\exp(-\alpha\rho_2)\left(-\alpha^2 + \frac{2\alpha-4}{\rho_1}\right) \\ &\quad \times \exp(-\alpha\rho_1)\exp(-\alpha\rho_2) \\ &= 16\pi^2 \int_0^\infty \rho_1^2\, d\rho_1 \exp(-2\alpha\rho_1)\left(-\alpha^2 + \frac{2\alpha-4}{\rho_1}\right) \\ &\quad \times \int_0^\infty \rho_2^2\, d\rho_2 \exp(-2\alpha\rho_2) \\ &= \pi^2(\alpha^{-4} - 4\alpha^{-5}).\end{aligned}$$

By symmetry, $\langle\psi_\alpha|\mathsf{K}_2|\psi_\alpha\rangle$ is the same.

The evaluation of

$$\langle\psi_\alpha|\mathsf{U}|\psi_\alpha\rangle = \int d^3\boldsymbol{\rho}_1 \int d^3\boldsymbol{\rho}_2 \frac{2}{|\boldsymbol{\rho}_1 - \boldsymbol{\rho}_2|}\exp(-2\alpha\rho_1)\exp(-2\alpha\rho_2)$$

is more awkward since the integral does not decompose into separate factors as the other ones do. However, we may use a result from gravitational potential theory which states that the gravitational potential of a *spherically symmetric* mass distribution at any external point is unaltered by concentrating all the mass at the centre. This allows us to replace $1/|\boldsymbol{\rho}_1 - \boldsymbol{\rho}_2|$ in the integrand by $1/\max(\rho_1, \rho_2)$, giving

$$\langle\psi_\alpha|\mathsf{U}|\psi_\alpha\rangle = 2\int \mathrm{d}^3\boldsymbol{\rho}_1 \int \mathrm{d}^3\boldsymbol{\rho}_2 \frac{\exp(-2\alpha\rho_1)\exp(-2\alpha\rho_2)}{\max(\rho_1, \rho_2)}$$

$$= 16\pi^2 . 4\int_0^\infty \rho_1^2\,\mathrm{d}\rho_1 \int_{\rho_1}^\infty \rho_2^2\,\mathrm{d}\rho_2 \frac{\exp(-2\alpha\rho_1)\exp(-2\alpha\rho_2)}{\rho_2}.$$

The last step follows by the symmetry of the integrand: the integration is performed only over $0 \leqslant \rho_1 \leqslant \rho_2 < \infty$, and doubled. The remainder of the work is straightforward, and gives

$$\langle\psi_\alpha|\mathsf{U}|\psi_\alpha\rangle = \tfrac{5}{4}\pi^2\alpha^{-5}.$$

Putting all these results together gives for the mean value to be minimized

$$\frac{\langle\psi_\alpha|(\mathsf{K}_1 + \mathsf{K}_2 + \mathsf{U})|\psi_\alpha\rangle}{\langle\psi_\alpha|\psi_\alpha\rangle} = \alpha^6(2\alpha^{-4} - 8\alpha^{-5} + \tfrac{5}{4}\alpha^{-5})$$

$$= 2\alpha^2 - \tfrac{27}{4}\alpha.$$

The minimum value occurs for $\alpha = \frac{27}{16}$ and is in fact $-\frac{729}{128} = -5{\cdot}69\ldots$. More refined calculations give a value of $-5{\cdot}81$.

The physical meaning of this result is that the minimum energy required to achieve the double ionization of helium (where both electrons are removed to infinity) is

$$\frac{\hbar^2}{2m_e a_0^2} \times 5{\cdot}81.$$

Double ionization may be regarded as a two-stage process:

$$\underset{(-5{\cdot}81)}{\mathrm{He}} \xrightarrow{1{\cdot}81} \underset{(-4)}{\mathrm{He}^+} \xrightarrow{4} \underset{(0)}{\mathrm{He}^{++}}.$$

Since the ground state energy of He^+ is -4 units (p. 102), the energy for single ionization is easily deduced, and is $1{\cdot}81$ units. These figures are in good agreement with experiment.

The Van der Waals attraction

The second part of this chapter will be devoted to a variational calculation of the force between two hydrogen atoms whose separation is large compared to a_0. A good way to calculate this force is to evaluate $E(R)$, the energy of the ground state of the entire system (not forgetting the mutual Coulomb interaction of the protons), as a function of R, the distance between the protons. The required force is then exactly $-\mathrm{d}E(R)/\mathrm{d}R$.

A qualitative idea of the shape of the function $E(R)$ is easily gained. If R is very small (the protons are very close), then the protonic repulsion dominates, and $E(R)$ is large and positive. As R increases, $E(R)$ goes negative, and eventually reaches a minimum: we know this has to happen, since we know that there exists a stable configuration—the hydrogen molecule. Beyond the minimum of $E(R)$, the force is attractive. As $R \to \infty$, $E(R)$ tails off to $2E_0$, where E_0 is the (negative) ground state energy of an isolated atom.

We shall be interested in the interaction when R is large compared to a_0. The effect may then be approximated as follows: Suppose that atom 1 is slightly distorted to acquire an electric dipole moment. The field of a dipole moment falls off as r^{-3}, and atom 2 experiences an electric intensity $\mathscr{E}$ proportional (among other things) to R^{-3}. The energy change of atom 2 in this field depends on $\mathscr{E}^2$ (p. 159), which goes as R^{-6}. Thus we may use the picture of a pair of induced dipole moments, each maintained by the electric field of the other, with a mutual potential energy proportional to R^{-6}. Of course, when the atoms get close enough to 'see' more structure than their overall dipole moments, this arguments begins to fail.

The purpose of the variational calculation is to estimate the *coefficient* of R^{-6} in this mutual potential energy.

The trial function

In the presence of an external electric field the ground state of a hydrogen atom is distorted. In choosing a suitable trial function it is important to have some idea of the nature of this distortion. Now it happens that we have already looked at this problem from the point of view of a second order perturbation (p. 159); the result was that the distorted ground-state wavefunction took the form

$$\psi(\boldsymbol{\rho}) = e^{-\rho} + \alpha(\rho + \tfrac{1}{2}\rho^2)e^{-\rho}\cos\theta \quad \text{(unnormalized)}$$

where the constant α was there proportional to the external field. Here, of course, the field is *not* external: it is generated by the mutual interaction of the two hydrogen atoms, and is not known to begin with. What is known is that ψ has the right kind of shape, and may therefore be used to build up a suitable trial function for a variational calculation.

In fact, it is necessary to generalize somewhat. We shall consider states for each hydrogen atom which are built up from four basic state vectors: one being the undistorted ground S-state

$$|\mathrm{s}\rangle \leftarrow \mathrm{e}^{-\rho},$$

and the others being the three P-state vectors which characterize the distortion in a small electric field

$$|\mathrm{x}\rangle \leftarrow (\rho+\tfrac{1}{2}\rho^2)\,\mathrm{e}^{-\rho}\sin\theta\cos\phi.$$
$$|\mathrm{y}\rangle \leftarrow (\rho+\tfrac{1}{2}\rho^2)\,\mathrm{e}^{-\rho}\sin\theta\sin\phi,$$
$$|\mathrm{z}\rangle \leftarrow (\rho+\tfrac{1}{2}\rho^2)\,\mathrm{e}^{-\rho}\cos\theta.$$

The general linear combination of these may be written

$$|\psi_{\alpha,\boldsymbol{u}}\rangle \leftarrow \psi(\boldsymbol{\rho};\alpha,\boldsymbol{u})$$
$$= \mathrm{e}^{-\rho}+\alpha(\rho+\tfrac{1}{2}\rho^2)\,\mathrm{e}^{-\rho}\boldsymbol{u}.(\sin\theta\cos\phi, \sin\theta\sin\phi, \cos\phi),$$

defined by the constant α and the unit vector $\boldsymbol{u}$.

Evaluating the integrals

As usual, there are many integrals to be evaluated; the work is not unlike that for the helium atom. First

$$\langle \mathrm{s}|\mathrm{s}\rangle = \int \mathrm{d}^3\boldsymbol{\rho}\,\mathrm{e}^{-2\rho} = 4\pi\int_0^\infty \rho^2\,\mathrm{d}\rho\,\mathrm{e}^{-2\rho} = \pi,$$

$$\langle \mathrm{z}|\mathrm{z}\rangle = \int \mathrm{d}^3\boldsymbol{\rho}(\rho+\tfrac{1}{2}\rho^2)^2\,\mathrm{e}^{-2\rho}\cos^2\theta$$
$$= \int_0^\infty \rho^2\,\mathrm{d}\rho(\rho+\tfrac{1}{2}\rho^2)\,\mathrm{e}^{-2\rho}.\int_0^\infty \sin\theta\,\mathrm{d}\theta.\cos^2\theta.2\pi$$
$$= \tfrac{43}{8}\pi.$$

By symmetry, $\langle \mathrm{x}|\mathrm{x}\rangle = \langle \mathrm{y}|\mathrm{y}\rangle = \frac{43}{8}\pi$ also. There are 16 scalar products which may be formed from pairs of the four basic vectors; the θ and ϕ integrations ensure that the remaining 12 are zero. It is convenient to arrange these results in a 4×4 table of values of $\langle \mathrm{A}|\mathrm{B}\rangle$ with A, B = s, x, y, z in turn—for clarity, zero entries will from now on be indicated by dots:

$\langle \mathrm{A}\vert\mathrm{B}\rangle$:	s	x	y	z
s	π	·	·	·
x	·	$\frac{43}{8}\pi$	·	·
y	·	·	$\frac{43}{8}\pi$	·
z	·	·	·	$\frac{43}{8}\pi$.

Also needed are the matrix elements $\langle \mathrm{A}|\mathrm{K}|\mathrm{B}\rangle$, where K is the Hamiltonian for the hydrogen atom,

$$\mathrm{K} \leftarrow -\nabla_\rho^2 - \frac{2}{\rho}.$$

Note first that

$$\left(-\nabla_\rho^2 - \frac{2}{\rho}\right) e^{-\rho} = -e^{-\rho}$$

and

$$\left(-\nabla_\rho^2 - \frac{2}{\rho}\right)(\rho + \tfrac{1}{2}\rho^2)\, e^{-\rho}\cos\theta = -(\rho + \tfrac{1}{2}\rho^2)\, e^{-\rho}\cos\theta + 2\rho\, e^{-\rho}\cos\theta$$
$$= (\rho - \tfrac{1}{2}\rho^2)\, e^{-\rho}\cos\theta.$$

(The first says that $e^{-\rho}$ is an eigenstate of K; the second follows from the result on p. 100.) Then, for example,

$$\langle \mathrm{z}|\mathrm{K}|\mathrm{z}\rangle = \int d^3\boldsymbol{\rho}(\rho + \tfrac{1}{2}\rho^2)\, e^{-\rho}\cos\theta \,.\, (\rho - \tfrac{1}{2}\rho^2)\, e^{-\rho}\cos\theta$$
$$= -\tfrac{7}{8}\pi,$$

after a little work. Other integrals are similarly evaluated, giving

$\langle \mathrm{A}\vert\mathrm{K}\vert\mathrm{B}\rangle$:	s	x	y	z
s	$-\pi$	·	·	·
x	·	$-\tfrac{7}{8}\pi$	·	·
y	·	·	$-\tfrac{7}{8}\pi$	·
z	·	·	·	$-\tfrac{7}{8}\pi$.

Finally, in order to construct the interaction between the dipole moments of the two atoms, we need to evaluate expressions like $\langle \mathrm{A}|\boldsymbol{r}|\mathrm{B}\rangle$. For example,

$$\langle \mathrm{s}|\mathbf{z}|\mathrm{z}\rangle = \int d^3\boldsymbol{\rho}\, e^{-\rho} \,.\, a_0\rho\cos\theta \,.\, (\rho + \tfrac{1}{2}\rho^2)\, e^{-\rho}\cos\theta$$
$$= a_0 \int_0^\infty \rho^2\, d\rho\, \rho(\rho + \tfrac{1}{2}\rho^2)\, e^{-2\rho} \,.\, \int_0^\infty \sin\theta\, d\theta \cos^2\theta \,.\, 2\pi$$
$$= \tfrac{9}{4}\pi a_0.$$

(We have seen the $\tfrac{9}{4}$ before, on p. 159.) Similarly,

$$\langle \mathrm{s}|\mathrm{x}|\mathrm{z}\rangle = 0 = \langle \mathrm{s}|\mathrm{y}|\mathrm{z}\rangle,$$

and therefore

$$\langle \mathrm{s}|\boldsymbol{r}|\mathrm{z}\rangle = \tfrac{9}{4}\pi a_0 \mathbf{k},$$

k being the unit vector in the z-direction.

Other evaluations lead to the table

$\langle \mathrm{A}\|\boldsymbol{r}\|\mathrm{B}\rangle$:	s	x	y	z
s	·	$\frac{9}{4}\pi a_0\mathbf{i}$	$\frac{9}{4}\pi a_0\mathbf{j}$	$\frac{9}{4}\pi a_0\mathbf{k}$
x	$\frac{9}{4}\pi a_0\mathbf{i}$	·	·	·
y	$\frac{9}{4}\pi a_0\mathbf{j}$	·	·	·
z	$\frac{9}{4}\pi a_0\mathbf{k}$	·	·	·

The three tables contain the integrals that we need for all the remaining work; there are 80 in total, of which all but 14 are zero. Really only five need to be evaluated from scratch, since the others may be deduced by geometrical symmetry arguments.

For the trial function $\psi_{\alpha u}$ (p. 212), it follows from these results that

$$\langle\psi_{\alpha u}|\psi_{\alpha u}\rangle = \pi(1+\tfrac{43}{8}\alpha^2),$$
$$\langle\psi_{\alpha u}|\mathsf{K}|\psi_{\alpha u}\rangle = -\pi(1+\tfrac{7}{8}\alpha^2),$$
$$\langle\psi_{\alpha u}|\boldsymbol{r}|\psi_{\alpha u}\rangle = \pi a_0\,.\,\tfrac{9}{2}\alpha\boldsymbol{u}.$$

PROBLEM

For a hydrogen atom in an external field $\mathscr{E}$ the Hamiltonian is

$$\mathsf{H} = -\frac{\hbar^2}{2m_e a_0^2}\mathsf{K}+e\mathscr{E}\,.\,\boldsymbol{r}.$$

Use the results of this section to show that, for the state $|\psi_{\alpha u}\rangle$, the mean value of H is

$$\langle H\rangle_{\alpha u} = \frac{\dfrac{\hbar^2}{2m_e a_0^2}(1+\tfrac{7}{8}\alpha^2)+\tfrac{9}{2}ea_0\alpha\mathscr{E}\,.\,\boldsymbol{u}}{(1+\tfrac{43}{8}\alpha^2)},$$

and that this is a minimum for

$$\alpha = \frac{2\eta}{1+(1+\frac{43}{2}\eta^2)^{\frac{1}{2}}},$$

where η is the dimensionless parameter $|ea_0\mathscr{E}|/(\hbar^2/m_e a_0^2)$. (Hint: first choose $\boldsymbol{u}$ to make the term containing $\mathscr{E}$ as negative as possible.)

Show that the minimum of $\langle H\rangle$ is

$$E_{\text{est}} = -\frac{\hbar^2}{2m_e a_0^2}(1+\tfrac{9}{2}\eta^2)+0(\eta^4),$$

and comment on the relationship to the polarizability calculated on p. 159.

The Van der Waals force: a first attempt

The nuclei of two hydrogen atoms are a distance R apart: how do the atoms interact? We assume that R is large enough for the dipole-dipole interaction to be valid as an approximation and therefore take as Hamiltonian

$$\mathsf{H} = \frac{\hbar^2}{2m_e a_0^2}(\mathsf{K}_1 + \mathsf{K}_2 + \mathsf{W})$$

where

$$\mathsf{K}_1 \leftarrow -\nabla_{\rho_1}^2 - \frac{2}{\rho_1}$$

is the Hamiltonian for the electron in atom 1; its position is $\boldsymbol{\rho}_1$ with respect to the nucleus. K_2 is similarly defined for atom 2. W is the mutual potential energy of the dipole moments μ_1, μ_2 of the two atoms, adopted from electrostatic theory (p. 168; the 'contact term' is irrelevant here),

$$\frac{\hbar^2}{2m_e a_0^2} W = \frac{1}{4\pi\varepsilon_0 R^3}\{\boldsymbol{\mu}_1 . \boldsymbol{\mu}_2 - 3(\mathbf{k} . \boldsymbol{\mu}_1)(\mathbf{k} . \boldsymbol{\mu}_2)\};$$

here the z-axis has been chosen to contain both protons. Now, since

$$\boldsymbol{\mu}_1 = -e\boldsymbol{r}_1 \quad \text{and} \quad \boldsymbol{\mu}_2 = -e\boldsymbol{r}_2$$

we may rearrange slightly to give

$$W = \frac{2a_0}{R^3}\{\boldsymbol{r}_1 . \boldsymbol{r}_2 - 3(\mathbf{k} . \boldsymbol{r}_1)(\mathbf{k} . \boldsymbol{r}_2)\}.$$

By the way, again we see how useful the Bohr radius a_0 is as a scaling parameter.

The choice of trial function must now be made. Our experience with the helium atom suggests that a reasonable choice might be a product of two functions, one for each of the two electrons, each of the right kind of shape:

$$\begin{aligned}\psi(\boldsymbol{\rho}_1, \boldsymbol{\rho}_2) &= \psi(\boldsymbol{\rho}_1; \alpha\boldsymbol{u}_1)\psi(\boldsymbol{\rho}_2; \alpha\boldsymbol{u}_2),\\ &= \psi_1\psi_2, \quad \text{say.}\end{aligned}$$

(Symmetry between the atoms requires that α should be the same for each factor.) In fact, this is quite disastrous, and for the sake of physical insight it is important to know why. So let us investigate the consequences of this choice.

Since ψ is a product of two one-electron functions, we may use the final results of the last section very straightforwardly. First, (see the comments on the direct product, p. 125),

$$\langle\psi|\psi\rangle = \langle\psi_1|\psi_1\rangle\langle\psi_2|\psi_2\rangle = \pi^2(1 + \tfrac{43}{8}\alpha^2)^2,$$

$$\begin{aligned}\langle\psi|\mathsf{K}_1|\psi\rangle &= \langle\psi_1|\mathsf{K}_1|\psi_1\rangle\langle\psi_2|\psi_2\rangle\\ &= -\pi^2(1 + \tfrac{7}{8}\alpha^2)(1 + \tfrac{43}{8}\alpha^2);\end{aligned}$$

$\langle\psi|\mathsf{K}_2|\psi\rangle$ is the same, by symmetry. Also

$$\langle\psi|\mathsf{W}|\psi\rangle = \frac{2a_0}{R^3}(\langle\psi_1|\boldsymbol{r}_1|\psi_1\rangle . \langle\psi_2|\boldsymbol{r}_2|\psi_2\rangle - 3\langle\psi_1|\mathbf{k}.\boldsymbol{r}_1|\psi_1\rangle\langle\psi_2|\mathbf{k}.\boldsymbol{r}_2|\psi_2\rangle)$$

$$= \pi^2 \tfrac{81}{2}\alpha^2\{\boldsymbol{u}_1 . \boldsymbol{u}_2 - 3(\mathbf{k}.\boldsymbol{u}_1)(\mathbf{k}.\boldsymbol{u}_2)\}\frac{a_0^3}{R^3},$$

The expression that is to be minimized is

$$\frac{\langle\psi|(\mathsf{K}_1+\mathsf{K}_2+\mathsf{W})|\psi\rangle}{\langle\psi|\psi\rangle} = -2\frac{1+\frac{7}{8}\alpha^2}{1+\frac{43}{8}\alpha^2}$$

$$+\tfrac{81}{2}\frac{\alpha^2}{(1+\frac{43}{8}\alpha^2)^2}\{\boldsymbol{u}_1 . \boldsymbol{u}_2 - 3(\mathbf{k}.\boldsymbol{u}_1)(\mathbf{k}.\boldsymbol{u}_2)\}\frac{a_0^3}{R^3}.$$

the parameters to be varied being α and the two unit vectors $\boldsymbol{u}_1$ and $\boldsymbol{u}_2$.

This expression is clearly to be as negative as we can make it. For a start, therefore, $\boldsymbol{u}_1$ and $\boldsymbol{u}_2$ must be chosen to make the expression between $\{\ldots\}$ as negative as possible. This will occur for $\boldsymbol{u}_1 = \boldsymbol{u}_2 = \pm\mathbf{k}$, the minimum value being -2. We are therefore led to choose α to minimize the expression

$$E(\alpha) = -2\frac{1+\frac{7}{8}\alpha^2}{1+\frac{43}{8}\alpha^2} - \frac{81\alpha^2}{(1+\frac{43}{8}\alpha^2)^2}\frac{a_0^3}{R^3}.$$

The usual technique leads to

$$\alpha = 0 \quad \text{or} \quad \pm\left(\tfrac{8}{43}.\frac{9a_0^3 - R^3}{9a_0^3 + R^3}\right)^{\frac{1}{2}},$$

for possible maxima and minima.

Now, if $R > 9^{1/3}a_0 = 2{\cdot}08a_0$, the only valid value for α is zero, and $E(\alpha) = -2$ precisely, independent of R: the predicted force is zero. If $R < 2{\cdot}08a_0$, then the true minimum is attained at one of the other values of α, but the hydrogen atoms are now so close that the dipole-dipole approximations fails completely anyway.

What has gone wrong? Simply, the trial function has the wrong shape! Each of the two factors is a linear combination of the four functions listed on p. 212; the product of these factors therefore inevitably includes contributions like, for example,

$$e^{-\rho_1} . (\rho^2 + \tfrac{1}{2}\rho_2^2)\, e^{-\rho_2}\cos\theta_2.$$

Such a term goes more harm than good. It is associated with a distortion of atom 2 with no corresponding distortion in atom 1. It therefore tends to increase the mean value $\langle K_2\rangle$, but fails to provide a compensating decrease in $\langle W\rangle$. Its presence will always make the trial function poorer. In fact, the inclusion of such terms makes the trial function so poor that, for $R > 2{\cdot}08a_0$, the best function is the undistorted one. In the figure, $E(\alpha)$ is plotted against α for three typical values of R/a_0, to illustrate the unhelpful behaviour of the minimum.

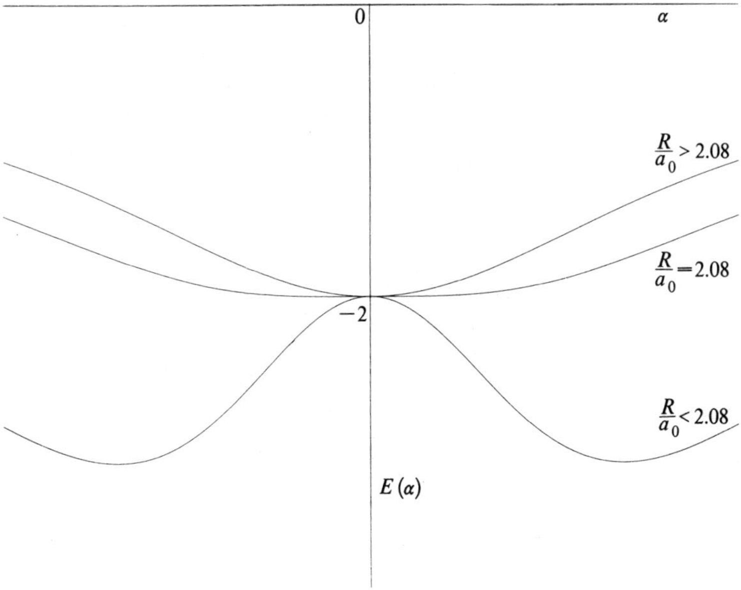

The moral is that, unless the trial function has been chosen to reflect the physics of the problem adequately, we have already failed.

The Van der Waals force: a good trial function

The cure for our problems is to make the trial function more flexible and thus, we hope, more realistic. Consider the combined states $|B_1B_2\rangle$, for which electron 1 is in state $B_1(= s, x, y$ or $z)$ and electron 2 is in state $B_2(= s, x, y, z)$. There are 16 such states in all; for example,

$$|sz\rangle \leftarrow e^{-\rho_1}.(\rho_2+\tfrac{1}{2}\rho_2^2)\,e^{-\rho_2}\cos\theta_2\,,$$

(cf. p. 212). We shall use as trial state a linear combination of all 16 statevectors

$$|\psi\rangle = \sum_{B_1B_2} |B_1B_2\rangle\alpha_{B_1B_2},$$

and vary the 16 αs to obtain the best fit. (We shall not be surprised to find that some of the αs are zero, since we expect to exclude certain contributions—$|sz\rangle$ is one, for example—by the arguments of the last section.)

The recipe of p. 205 applies here, and leads us to evaluate the 16×16 array of values

$$S_{A_1A_2,B_1B_2} = \langle A_1A_2|(K_1+K_2+W)|B_1B_2\rangle - E\langle A_1A_2|B_1B_2\rangle$$

for all 256 combinations of A_1, A_2, B_1, B_2 chosen from s, x, y, or z. Because the basic wavefunctions are all products of single-electron wavefunctions, a typical element may be decomposed in the following way:

$$\begin{aligned} S_{A_1A_2,B_1B_2} &= \langle A_1|K_1|B_1\rangle\langle A_2|B_2\rangle + \langle A_1|B_1\rangle\langle A_2|K_2|B_2\rangle \\ &\quad + \frac{2a_0}{R^3}\big(\langle A_1|\boldsymbol{r}_1|B_1\rangle . \langle A_2|\boldsymbol{r}_2|B_2\rangle \\ &\quad - 3\langle A_1|\mathbf{k}.\boldsymbol{r}_1|B_1\rangle\langle A_2|\mathbf{k}.\boldsymbol{r}_2|B_2\rangle\big) \\ &\quad - E\langle A_1|B_1\rangle\langle A_2|B_2\rangle. \end{aligned}$$

Every one of the expressions on the right has been evaluated already (pp. 212–14), and all that is now needed is the careful assembly of the 256 possibilities into the 16×16 matrix on the next page. For example, the non-zero terms in $S_{sx,sx}$ are

$$\begin{aligned} &\langle s|K_1|s\rangle\langle x|x\rangle + \langle s|s\rangle\langle x|K_2|x\rangle - E\langle s|s\rangle\langle x|x\rangle \\ &= -\pi . \tfrac{43}{8}\pi - \pi . \tfrac{7}{8}\pi - E\pi . \tfrac{43}{8}\pi \\ &= \pi^2(-\tfrac{25}{4} - \tfrac{43}{8}E). \end{aligned}$$

It turns out that all but 28 entries are zero, and that the non-zero ones take one of only five different possible values.

The recipe declares that the value of E that we want is the lowest that will make the 16×16 determinant zero. The determinant is easily evaluated, since it can be put in a block diagonal form by suitable permutations of the rows and columns, yielding a product of much smaller determinants:

$$\begin{vmatrix} d_1 & c_1 & c_1 & c_2 \\ c_1 & d_3 & \cdot & \cdot \\ c_1 & \cdot & d_3 & \cdot \\ c_2 & \cdot & \cdot & d_3 \end{vmatrix} \times \begin{vmatrix} d_2 & c_1 \\ c_1 & d_2 \end{vmatrix}^2 \times \begin{vmatrix} d_2 & c_2 \\ c_2 & d_2 \end{vmatrix} \times |d_3|^6 = 0$$

The E we seek must make one of these factors zero: which one? As the undistorted state for $R = \infty$ is the ss-state, the relevant factor is the one which contains the (ss, ss)-element, namely d_1. Thus we require

$$\begin{vmatrix} d_1 & c_1 & c_1 & c_2 \\ c_1 & d_3 & \cdot & \cdot \\ c_1 & \cdot & d_3 & \cdot \\ c_2 & \cdot & \cdot & d_3 \end{vmatrix} \equiv d_3^2(d_1d_3 - 2c_1^2 - c_2^2) = 0.$$

Now $d_3 = 0$ gives $E = -14/43$, independent of R. It is therefore the other factor which is to be zero. Inserting values for d_1, d_3, c_1 and c_2, and rearranging gives a quadratic equation for E,

$$1849E^2 + 4300E + 1204 - 39\,366a_0^6R^{-6} = 0$$

The secular determinant $S_{A_1A_2,B_1B_2}$

	ss	sx	sy	sz	xs	ys	zs	xx	xy	xz	yx	yy	yz	zx	zy	zz
ss	d_1	·	·	·	·	·	·	c_1	·	·	·	c_1	·	·	·	c_2
sx	·	d_2	·	·	c_1	·	·	·	·	·	·	·	·	·	·	·
sy	·	·	d_2	·	·	c_1	·	·	·	·	·	·	·	·	·	·
sz	·	·	·	d_2	·	·	c_2	·	·	·	·	·	·	·	·	·
xs	·	c_1	·	·	d_2	·	·	·	·	·	·	·	·	·	·	·
ys	·	·	c_1	·	·	d_2	·	·	·	·	·	·	·	·	·	·
zs	·	·	·	c_2	·	·	d_2	·	·	·	·	·	·	·	·	·
xx	c_1	·	·	·	·	·	·	d_3	·	·	·	·	·	·	·	·
xy	·	·	·	·	·	·	·	·	d_3	·	·	·	·	·	·	·
xz	·	·	·	·	·	·	·	·	·	d_3	·	·	·	·	·	·
yx	·	·	·	·	·	·	·	·	·	·	d_3	·	·	·	·	·
yy	c_1	·	·	·	·	·	·	·	·	·	·	d_3	·	·	·	·
yz	·	·	·	·	·	·	·	·	·	·	·	·	d_3	·	·	·
zx	·	·	·	·	·	·	·	·	·	·	·	·	·	d_3	·	·
zy	·	·	·	·	·	·	·	·	·	·	·	·	·	·	d_3	·
zz	c_2	·	·	·	·	·	·	·	·	·	·	·	·	·	·	d_3

Key: $d_1 = \pi^2(-2 - E)$

$d_2 = \pi^2(-\frac{25}{4} - \frac{43}{8}E)$

$d_3 = \pi^2(-\frac{301}{32} - \frac{1849}{64}E)$

$c_1 = \pi^2(\frac{81}{8}a_0^3R^{-3})$

$c_2 = -2c_1$

whose lower root is

$$E_{\text{est}} = -(100+72(1+\tfrac{243}{8}a_0^6R^{-6})^{\frac{1}{2}})/86$$
$$= -2-12{\cdot}715\ldots\frac{a_0^6}{R^6}+\ldots.$$

This concludes the search. The joint ground-state energy of two hydrogen atoms for large R is dominated by

$$\frac{\hbar^2}{2ma_0^2}E_{\text{est}} = -\frac{\hbar^2}{ma_0^2}-6{\cdot}357\ldots\frac{\hbar^2a_0^4}{mR^6}.$$

To be wise after the event: it would have been possible to exclude states like $|\text{sx}\rangle$ from the start, in view of the comments in the last section. Also states like $|\text{xy}\rangle$ or $|\text{xz}\rangle$ may be excluded, as their contribution to $\langle W\rangle$ would change sign on reversing the direction of the x-axis. But the original problem is symmetric against such a reversal. This leaves only the four states $|\text{ss}\rangle$, $|\text{xx}\rangle$, $|\text{yy}\rangle$ and $|\text{zz}\rangle$ as possible contributors, and we could well have used only these four as our starting point. The result would have been the same.

The correct value of the coefficient a_0^6/R^6 is very nearly 13. An error as small as 2 per cent or so is really quite good for this kind of work.

18. Experience is the enemy of intuition

How obvious is quantum mechanics?

THE modern version of quantum mechanics appeared some fifty years ago, after a long period of groping and false starts. Since then it has served us well. Its rules are precise and unambiguous and, in every case where we can be sure that it has been correctly applied, it has been in superb agreement with experiment. Most physicists are therefore content.

However, it did not take long for certain 'philosophical' doubts to appear, mainly on account of the unavoidably statistical nature of the theory: only rarely is the result of an observation definitely predicted, since probabilities are usually all that we are given. Theorists whose prejudice inclined towards determinism were troubled, and they began to ask what was 'really' happening behind the scenes. Quantum mechanics not only gave no answer, but even declared that such questions ought not to be asked! Henceforth, intuition was to be regarded as a very fallible guide, and ideas which used to be regarded as obvious were to be viewed with reserve until experience either vindicated or invalidated them.

In this chapter I hope to be able to give a little of the flavour of the kind of conundrum that has arisen. There are two main ones, both with the air of paradox: one concerns the behaviour of a quantum particle between observations, and the other the curious correlations that may exist between apparently completely independent systems.

What is a quantum particle?

A frequently-given answer is 'sometimes a particle, sometimes a wave'. When I hear this answer, I like to add: 'and sometimes a clock'. However, the proper answer is 'none of these': to think of an electron as a *classical* particle or a *classical* wave, or even as some kind of paradoxical mixture of the two, is thoroughly misleading. The best description we know how to give is through the Schrödinger equation and the usual rules of quantum mechanics. It may be *convenient* to think of 'it' as a particle if we happen to be measuring its position $\boldsymbol{r}$, or as a wave with wavevector $\boldsymbol{k}$ if we happen to be measuring its momentum $\hbar\boldsymbol{k}$—or for that matter as a clock with frequency ν if we are measuring its energy $h\nu$ (precisely the example of the 'atomic clock'). (These are the various pictures associated with the r-, p-, and H-representations respectively. Now, what picture goes with the rlm-representation (p. 124)? Simply: 'particle-like' in the radial direction, and 'wave-like' in the $\theta\phi$-directions. Is it now clear how very much the discussion depends on what *we* happen to be interested in at the moment?)

Cloud-chamber tracks

When a fast-moving charged particle passes through a gas, it leaves a track of ionization in its wake. If the conditions are right, the ionized molecules in the track may be used as nucleation centres for the condensation of water vapour or the like; in this way the track is made visible. This is the principle of the **Wilson cloud-chamber**, used since 1912 to obtain the trajectories of particles taking part in nuclear reactions. A more modern apparatus is the **bubble chamber**, filled with slightly superheated liquid hydrogen. The local heating which results from the passage of a particle initiates boiling along the track, which is thus marked out by a trail of small bubbles.

Imagine now a cloud-chamber in the presence of a quantum particle with a nearly definite momentum $\hbar\boldsymbol{k}$; the wavefunction of the particle is nearly a plane wave

$$\psi(\boldsymbol{r}, t) = \text{const.} \exp(\mathrm{i}\boldsymbol{k}.\boldsymbol{r} - \mathrm{i}\omega t), \quad \text{with} \quad \hbar\omega = \hbar^2 k^2/2m.$$

(We have to say 'nearly' as we are thinking of *one* particle, and we ought strictly to use a normalizable ψ.) Hence the probability density $|\psi|^2$ is nearly uniform all over the cloud-chamber, and intuitively we would expect to see, not a track, but rather a general cloudiness throughout the gas.

This would be quite wrong, and would arise from a disastrous misunderstanding of quantum mechanics, which predicts that a track is always formed, whatever the shape of ψ may be.

It is convenient (the diagrams are easier) to consider a simpler one-dimensional case. A particle moves along the x-axis, on which there are two detectors A and B, either or both of which may be tripped by the passage of the particle. We shall assume that the detectors are delicate enough not to disturb the momentum of the particle overmuch. The only sure way of applying quantum mechanics to such a situation is to *include both detectors* in the system, which then becomes the direct product (p. 125) of one particle and two detectors. We shall consequently need *four* wavefunctions:

$\psi(x)$ for when neither A nor B is tripped

$\psi_{\mathrm{A}}(x)$ for when A alone is tripped,

$\psi_{\mathrm{B}}(x)$ for when B alone is tripped,

$\psi_{\mathrm{AB}}(x)$ for when both A and B are tripped.

When the particle is near $x = x_{\mathrm{A}}$, then A may get tripped; thus the

Hamiltonian—whatever it is—must couple $\psi(x_A)$ to $\psi_A(x_A)$, and also $\psi_B(x_A)$ to $\psi_{AB}(x_A)$. Similar couplings happen near $x = x_B$, on account of the detector B.

The initial state is

$$\psi(x) = e^{ikx}, \quad \psi_A(x) = \psi_B(x) = \psi_{AB}(x) = 0,$$

that is, an undetected particle with momentum $\hbar k$ (diagram (a)). Shortly afterwards, both ψ_A and ψ_B are non-zero for a short distance 'downstream', on account of the couplings (diagram (b)). But ψ_{AB} is still zero, and *must*

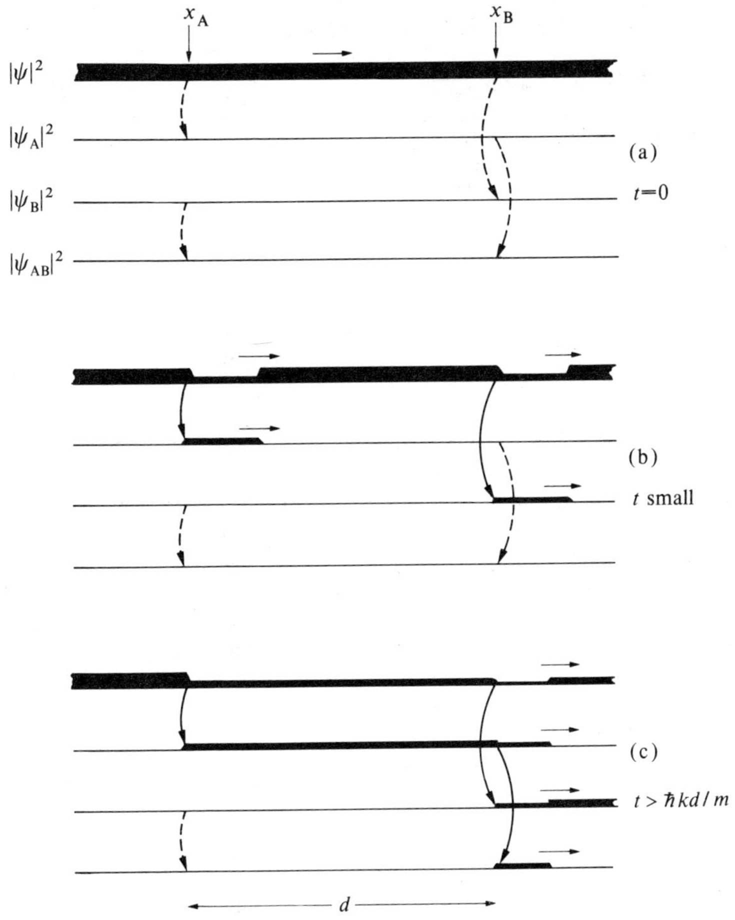

remain zero until the front of the disturbance in ψ_A passes x_B (diagram (c)). The velocity of this front is the *group velocity* of the wave, $d\omega/dk = \hbar k/m$, and this is exactly the velocity of the particle in the classical sense. Thus, even though either A or B may be tripped almost immediately, they cannot both be tripped until enough time for a classical particle to get from A to B has passed, and B is then always the last to go.

In three dimensions the conditions for both A and B to be tripped are more stringent. As time passes, the region within which ψ_A is non-negligible trails out into a narrow cone with vertex at A; similarly for ψ_B. The

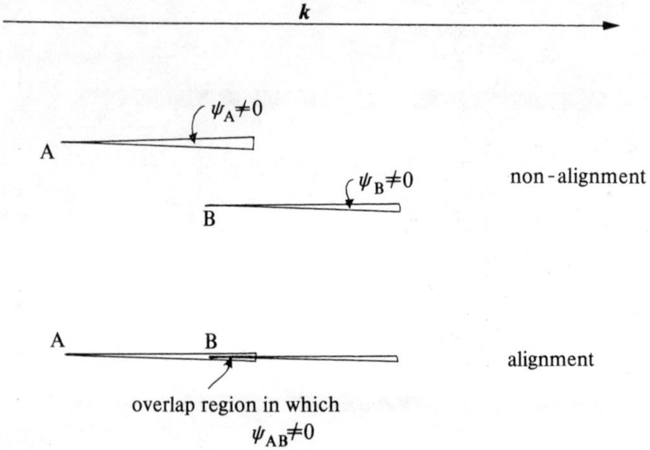

component ψ_{AB} will have no chance of ever being non-negligible unless the cone trailing from A ultimately contains B (or vice versa); thus before both detectors can be tripped not only must a certain time elapse but also the line AB must be very nearly parallel to the propagation vector $\boldsymbol{k}$.

So much for two detectors. It should be clear that the same kind of qualitative argument will go through for the much larger number N of 'detectors' which make up the gas in a Wilson cloud chamber: the conclusion is that the detectors which are tripped by the particle must all lie close to a single line parallel to $\boldsymbol{k}$, and this is exactly what we mean by a 'track'. Conventional quantum mechanics thus does not always allow us to think of the electron as just a single simple wave, though it should be said in passing that there are continuing attempts to formulate alternative theories where this could be nearer the truth. (To do the thing properly, we require the distinct functions ψ, ψ_A, ψ_B, ..., ψ_{AB}, ..., ψ_{ABC}, ..., numbering in all 2^N; for any ordinary cloud chamber this number is unimaginably huge, incomparably greater for example than the number of electrons in the observable Universe!)

Fresnel's interference experiment

At the beginning of the nineteenth century it was becoming understood that visible light is a wave motion of some kind, mainly on account of diffraction and interference phenomena. A celebrated experiment to demonstrate interference was set up by Thomas Young; a variant proposed by Fresnel is shown in the diagram. Monochromatic light from a single

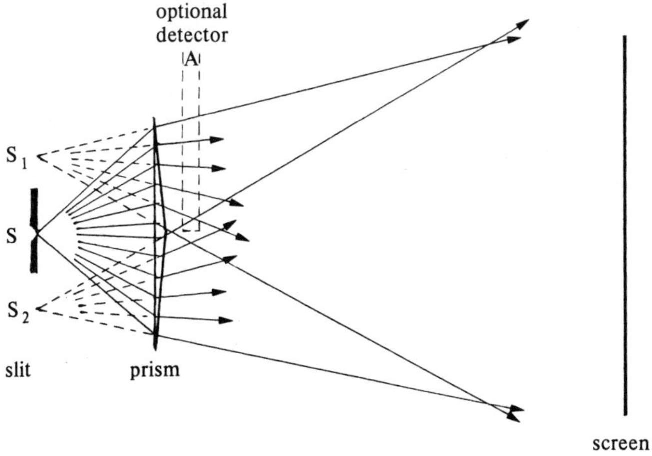

narrow slit source S can pass through either half of a very obtuse glass prism, and then falls on a screen some distance away, where interference fringes are seen. That the fringes are due to *interference* between the two images S_1 and S_2 of the slit is clearly shown by masking off one half of the prism: the *average* brightness at the screen must obviously drop, but at places where destructive interference previously occurred the screen actually becomes *brighter*. Such behaviour is inconceivable under any ordinary corpuscular theory of light, and it was not long before the wave theory was universally accepted.

Nowadays, however, we also 'believe' in photons; indeed, it is possible to observe individual photons with the help of a photomultiplier or an image intensifier, which are simply means of pushing the photoelectric effect to its technological limit. Imagine the light from the slit dimmed to such a low level that there is at any time only one photon on average between the slit and the screen, and mark the position of impact of each photon as it arrives at the screen. Quantum mechanics predicts that the same interference pattern is *still there*, now being built up in a statistical manner by an accumulation of successive impacts; the result looks like a more or less grainy photograph, as shown.

Now imagine the upper half of the prism to be masked off. Since the screen becomes brighter at certain places, we have to agree that *more* (not fewer) photons arrive at these places than before. If the photon were a particle in the usual sense of this word, we would like to be able to say that any particular one has passed *either* through the upper half of the prism *or* through the lower half, and it would then be inconceivable that the screen should be brighter anywhere when the mask is in place. Clearly, therefore, we must refrain from ascribing a trajectory to the photon.

But there is really no reason to refrain from asking: 'which half of the prism has the photon traversed?' All we need to do is to insert a perfectly transparent photon detector A behind the upper half of the prism, and observe its response (diagram, p. 225). (Actually, such a detector is impossible to come by, but it does no harm to imagine one.) Quantum theory gives a clear prediction of the result. Suppose that in the absence of the detector the photon wavefunction is ψ_0 (in fact, it is a solution of Maxwell's equations, but this need not detain us), and that

$$\psi_0 = \psi_U + \psi_L$$

where ψ_U and ψ_L are respectively the contributions from the upper and lower half of the prism. When the detector A is included, the state for the entire setup will require *two* wavefunctions ψ and ψ_A ('the detector has not/has tripped'). Let us presume that now, to the right of the detector,

$$\psi = \alpha\psi_U + \psi_L,$$

$$\psi_A = \beta\psi_U.$$

If $\alpha = 0$ and $\beta = 1$, the detector is extremely efficient, 'transferring' all of ψ_U from ψ to ψ_A. If $\alpha = 1$ and $\beta = 0$, the detector is completely ineffective. There is a continuous range of possibilities between these two extremes. (In real life, any detector will have a much more drastic effect on ψ_U than merely to transfer some of it from ψ to ψ_A, but for our qualitative discussion this need not worry us.)

The interference pattern is predicted by calculating the probability density for finding the photon at a particular place on the screen, whether the detector has been tripped or not; this is

$$|\psi|^2 + |\psi_A|^2 = |\alpha\psi_U + \psi_L|^2 + |\beta\psi_U|^2.$$

For $\alpha = 1$ (the detector being ineffective) this is the original $|\psi_0|^2$, and the interference pattern is perfect. For $\alpha = 0$ (the completely efficient detector) the probability density is $|\psi_U|^2 + |\psi_L|^2$, and is the *straight sum* of the

screen
S1
S2

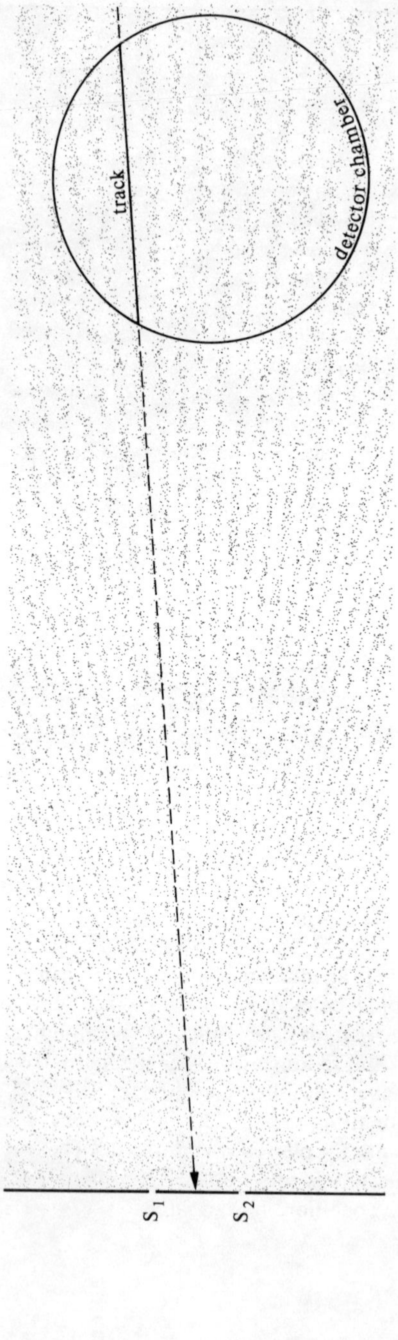
track
detector chamber
S_1
S_2

separate intensities from the two slits; the interference pattern is entirely absent. How strongly the pattern is present clearly depends on how inefficient the detector is: the *more certain our knowledge* about which half of the prism has been encountered by the photon, the *weaker* the interference. (Put like this, it sounds very subjective, but that is the way things appear to be.)

A possible objection is that the detector has been put in the wrong place, and that it would be an improvement to give the interference pattern a better chance of being built up. Let us therefore remove the detector A for good, and replace the *screen* by a closely-spaced array of photon detectors, designed to track the *ultimate* trajectory of the photon—a kind of 'photon cloud chamber', if such a thing were possible. The idea is then to produce the trajectories backwards to determine which slit image they emanate from. But this won't work, since it may be shown that the most probable trajectories run along the ridges of maximum amplitude of ψ; these ridges form a system of confocal hyperbolas whose asymptotes pass neatly between the two slit images, leaving the question unresolved once again.

The reader is probably now ready to accept that, far enough 'downstream', all information about which slit any photon has come from is entirely lost. He will be quite wrong, however! Replace the original

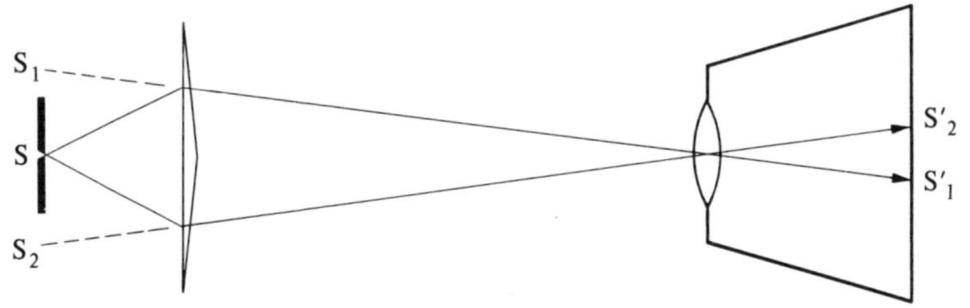

screen by a *camera*, focussed on the pair of slit images. It is perfectly clear that we can now tell at which of S_1 or S_2 a photon originates by observing whether it is recorded at S'_1 or S'_2. On the other hand, we now have no chance of observing the interference pattern: the camera is in the way!

The moral should be evident: quantum mechanics describes 'physical reality'—whatever *that* may be—rather well, certainly at the level we have been considering, and so long as we keep within the 'rules', we shall not go far wrong. However, we shall do well to keep a tight hold on our imagination, recognizing the pitfalls of questions like: 'What is *really* happening? How has the photon *really* moved?' The only legitimate way to answer such a question is to set up an appropriate apparatus (if possible) which will 'ask' the question. In so doing, however, we may be called upon to undertake a substantial reformulation of the problem. We have already seen, for example, how the proper theory of a cloud-chamber track will not allow us to get away with the simple electron wavefunction that we may have hoped for to begin with.

Probing the hydrogen atom

The interference pattern in the two-slit experiment is a relatively large-scale phenomenon. However, there is no reason to suppose that the same kind of effect will not appear at much shorter distances. For example, what does the electron in a hydrogen atom do?

To answer such a question, we must propose a definite experimental procedure; for example, let us make an *accurate* measurement of the position of the electron relative to the nucleus, and plot the result in the obvious way as a dot on a (three-dimensional) diagram. So far we have learned very little. But now repeat the procedure for a long sequence of hydrogen atoms, all in the same quantum state; as the number of dots grows, their density in any region becomes more or less proportional to

(a)

$|\psi|^2$, where ψ is the wavefunction for the state concerned. The figure gives some computer-generated impressions of a few such diagrams, or, more exactly, of a thin slice containing the origin; in each case, the dotted circle has radius a_0, the Bohr radius, and is included for comparison. Figures (a) and (b) are both spherically symmetric ($l = 0$), and illustrate the ground state and the first excited state respectively. Figure (c) illustrates an $l = 1$ state, while (d) exhibits the lopsidedness of the ground state in the presence of a uniform constant electric field—the second-order Stark effect, in fact (p. 159).

This way of building up a picture of the state of the atom clearly demonstrates the statistical nature of quantum mechanics, and is in line with the discussion on p. 6. We have to be careful at this point: we are not building up a picture of one atom in a particular state, but *a composite overlap of many*. In fact, an observation of the kind we have in mind is very violent: to determine the electron position accurately enough requires a

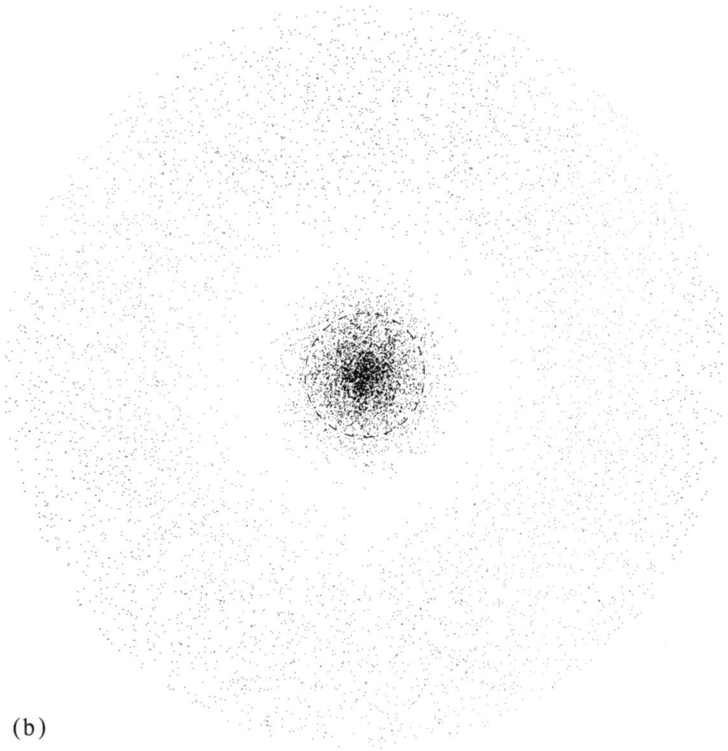

(b)

very fine probe, namely, a quantum particle of very small wavelength, hence of large momentum, and hence of high energy—high enough, in fact, inevitably to ionize the atom at the moment of observation. We are promised only one bite at the cherry!

The paradox of correlation

One of the more surprising and enthralling new features of quantum mechanics was that a composite system is incomparably more than the mere sum of its parts. One consequence is practical: if we apply the simple methods which work well for, say, a hydrogen or helium atom to molecules of only moderate complexity, we shall be faced with numerical labour of astronomical proportions. However, this is not our main concern here. The most disturbing consequences are philosophical in nature. The classical one was formulated in the mid-thirties by Einstein, Podolsky and Rosen;

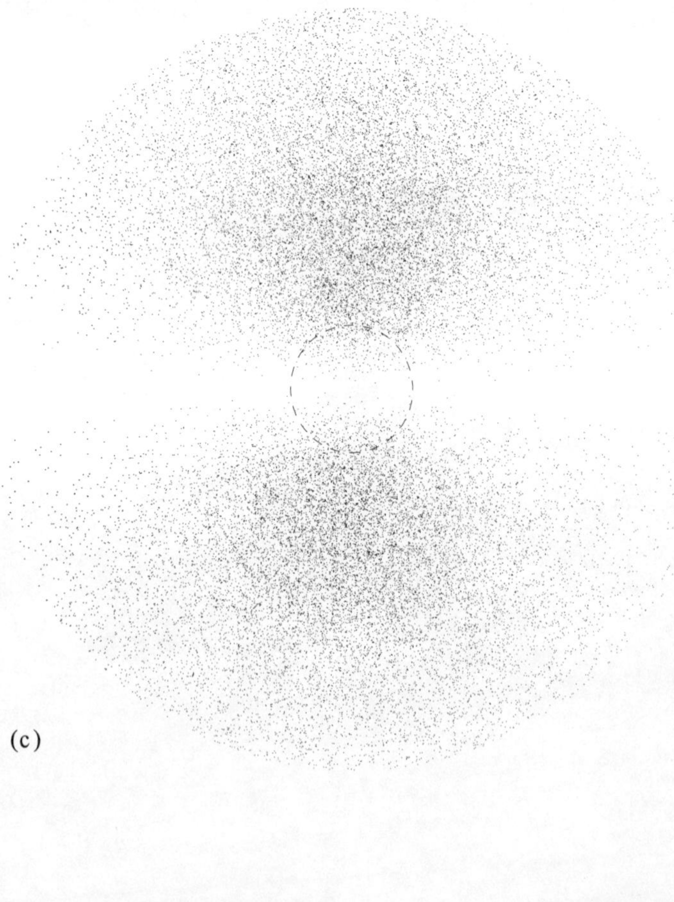

(c)

roughly put, once two systems have been in contact—no matter how long ago, and no matter how far apart they may be now—observing one system *drastically alters* our view of the other in ways which are difficult to understand. Such correlations are clearly and unambiguously predicted by conventional quantum mechanics; experimental evidence is in their favour; and yet we are unable to comprehend how they can happen.

There is nothing wrong with correlation as such: life is full of it. A universe without correlation would be chaotic, uneventful and dull. But quantum mechanics appears to yield *too much* correlation, more than we can intuitively stomach without feeling compelled to ask what is really going on.

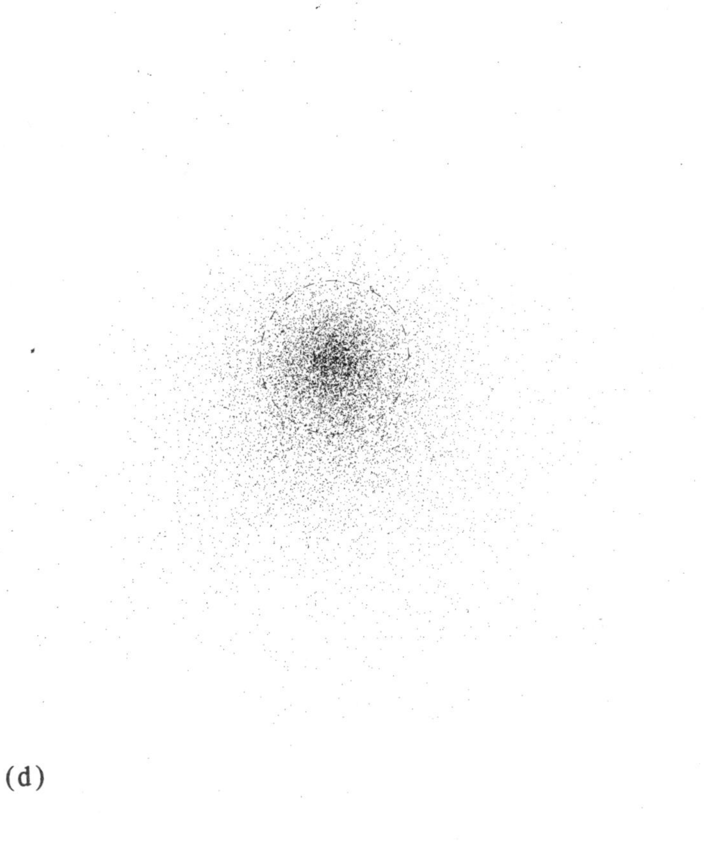

(d)

A simple example illustrates the problem in a very few lines. Indeed, much of the work is already to be found on page 170, where there is given a 4×4 matrix representation of the six spin components $\boldsymbol{\sigma}_e$ and $\boldsymbol{\sigma}_p$ of two spin-$\frac{1}{2}$ particles—electron and proton, in fact. Now think about some curious properties of the joint state S of the two spins, represented by a column vector

$$S \leftarrow 2^{-\frac{1}{2}} \begin{pmatrix} 0 \\ 1 \\ -1 \\ 0 \end{pmatrix}.$$

(In the context of pp. 169–71, this is the singlet level in the hyperfine splitting of the hydrogen atom ground state, with zero total angular momentum.)

It is simple to verify that, for the state S, the mean value of each component of each spin is zero,

$$\langle \boldsymbol{\sigma}_e \rangle_S = 0, \quad \langle \boldsymbol{\sigma}_p \rangle_S = 0.$$

This is not at all remarkable as, being a singlet with zero total angular momentum, the state S has no preferred direction in space. It follows that measuring the component of electron spin *in any direction whatsoever* will give $\frac{1}{2}\hbar$ or $-\frac{1}{2}\hbar$ with equal probability: the electron thus gives the impression of being completely unpolarized. The same is true of the proton.

But what if we observe *both* spins, as in principle we are certainly entitled to do? Specifically, let us measure first σ_{pz}, and then σ_{ez}, in the state S. In connection with measuring σ_{pz} the formalism requires us to express the statevector of S as the unique superposition of eigenvectors of σ_{pz},

$$S \leftarrow 2^{-\frac{1}{2}} \begin{pmatrix} 0 \\ 1 \\ -1 \\ 0 \end{pmatrix} = 2^{-\frac{1}{2}} \begin{pmatrix} 0 \\ 1 \\ 0 \\ 0 \end{pmatrix} + 2^{-\frac{1}{2}} \begin{pmatrix} 0 \\ 0 \\ -1 \\ 0 \end{pmatrix}$$

[eigenvalue of σ_{pz}: -1 $+1$]

Having measured σ_{pz} and obtained (say) the value -1, we may then assert that the system is in the state

$$S_{-1} \leftarrow \begin{pmatrix} 0 \\ 1 \\ 0 \\ 0 \end{pmatrix},$$

in accord with the usual principles of quantum mechanics. Measuring σ_{ez} for the *electron* will now give $+1$ *with certainty*, as the statevector for S_{-1} is an eigenvector of σ_{ez}, in addition to σ_{pz}. Thus *the electron acquires a precise polarization in the z-direction as soon as the spin of the proton in the z-direction is investigated.*

One's reaction may well be: so what? In the hydrogen atom the electron and proton are very closely linked; surely looking at one will have an unavoidable influence on the other? The point here is that neither theory nor experiment appears to prohibit us from prising the electron and proton apart without affecting the spin states, *before* any measurements are made. (In practice, the procedure is a delicate one, but experiments on these lines are being planned for suitable systems.) Suppose then that the electron and proton, jointly in the state S, are separated by a considerable distance, a few metres perhaps, so that each may be examined without affecting the other. The quantum paradox is that examining either *must* affect the other: measuring the z-component of the proton spin leads to an immediate polarization of the electron spin in the z-direction, and how on earth does the electron, now at the far end of the laboratory, 'know' that the z-direction, and not some other direction, is the right one to be polarized in?

There is a deep and subtle problem here, and no-one has yet provided a satisfying resolution. One proposal is to assert that, *on separation*, each particle goes into its own private polarization state, the two states being oppositely oriented along a randomly selected direction. There are two objections: first, the theory provides no means whereby such a thing can happen—a Schrödinger equation will certainly not do it, and the necessary changes would be very deep-seated, and would very likely wreck the extraordinarily good correspondence between theory and experiment which we now enjoy. Second, even if an acceptable mechanism were found, the amount of correlation would almost always have to be less than that predicted by standard quantum mechanics, on account of **Bell's inequalities** which need to be satisfied if our natural intuitions are right. (That may sound very subjective; so it is!)

Conventional quantum mechanics violates Bell's inequalities in theory, and there is consequently a number of experiments designed to test if they are violated in real life. On balance, available results suggest that violation does indeed occur. It is therefore inevitable, for reasons of philosophical prejudice, that attempts to rebuild quantum mechanics will be made, to make any violation appear more 'reasonable'. Conceivably, however, we may have to give up the riddle, and say with Hamlet that all this 'should teach us there's a divinity that shapes our ends, rough-hew them how we will.'

Appendix: standard integrals

$$\int_0^\infty x^n\,\mathrm{e}^{-\lambda x}\,\mathrm{d}x = n!/\lambda^{n+1}$$

$$\int_0^\infty x^{2n}\,\mathrm{e}^{-\lambda x^2}\,\mathrm{d}x = \frac{\pi^{\frac{1}{2}}(2n)!}{2^{2n+1}n!\lambda^{n+\frac{1}{2}}}$$

$$\int_0^\infty x^{2n+1}\,\mathrm{e}^{-\lambda x^2}\,\mathrm{d}x = n!/2\lambda^{n+1}$$

Index

In the main text, bold type indicates the point at which a definition of a concept is given. When the index refers to any such point, the page number is given in bold type.

Physical constants and conversion factors

Avogadro constant	L or N_A	$6{\cdot}022 \times 10^{23}$ mol^{-1}
Bohr magneton	μ_B	$9{\cdot}724 \times 10^{-24}$ J T^{-1}
Bohr radius	a_0	$5{\cdot}292 \times 10^{-11}$ m
Boltzmann constant	k	$1{\cdot}381 \times 10^{-23}$ J K^{-1}
charge of an electron	e	$-1{\cdot}602 \times 10^{-19}$ C
Compton wavelength of electron	$\lambda_C = h/m_e c =$	$2{\cdot}426 \times 10^{-12}$ m
Faraday constant	F	$9{\cdot}649 \times 10^{4}$ C mol^{-1}
fine structure constant	$\alpha = \mu_0 e^2 c/2h =$	$7{\cdot}297 \times 10^{-3}$ ($\alpha^{-1} = 137{\cdot}0$)
gas constant	R	$8{\cdot}314$ J K^{-1} mol^{-1}
gravitational constant	G	$6{\cdot}673 \times 10^{-11}$ N m^2 kg^{-2}
nuclear magneton	μ_N	$5{\cdot}051 \times 10^{-27}$ J T^{-1}
permeability of a vacuum	μ_0	$4\pi \times 10^{-7}$ H m^{-1} exactly
permittivity of a vacuum	ε_0	$8{\cdot}854 \times 10^{-12}$ F m^{-1} ($1/4\pi\varepsilon_0 = 8{\cdot}988 \times 10^{9}$ m F^{-1})
Planck constant	h	$6{\cdot}626 \times 10^{-34}$ J s
(Planck constant)/2π	$\hbar$	$1{\cdot}055 \times 10^{-34}$ J s $= 6{\cdot}582 \times 10^{-16}$ eV s
rest mass of electon	m_e	$9{\cdot}110 \times 10^{-31}$ kg $= 0{\cdot}511$ MeV/c^2
rest mass of proton	m_p	$1{\cdot}673 \times 10^{-27}$ kg $= 938{\cdot}3$ MeV/c^2
Rydberg constant	$R_\infty = \mu_0^2 m_e e^4 c^3/8h^3 =$	$1{\cdot}097 \times 10^{7}$ m^{-1}
speed of light in a vacuum	c	$2{\cdot}988 \times 10$ m s^{-1}
Stefan–Boltzmann constant	$\sigma = 2\pi^5 k^4/15h^3c^2 =$	$5{\cdot}670 \times 10^{-8}$ W m^{-2} K^{-4}
unified atomic mass unit (^{12}C)	u	$1{\cdot}661 \times 10^{-27}$ kg $= 931{\cdot}5$ MeV/c^2
wavelength of a 1 eV photon		$1{\cdot}243 \times 10^{-6}$ m

1Å $= 10^{-10}$ m; 1 dyne $= 10^{-5}$ N; 1 gauss (G) $= 10^{-4}$ tesla (T);
0 °C $= 273{\cdot}15$ K; 1 curie (Ci) $= 3{\cdot}7 \times 10^{10}$ s^{-1};
1 J $= 10^7$ erg $= 6{\cdot}241 \times 10^{18}$ eV; 1 eV $= 1{\cdot}602 \times 10^{-19}$ J; 1 cal$_{th}$ $= 4{\cdot}184$ J;
ln 10 $= 2{\cdot}303$; ln $x = 2{\cdot}303$ log x; e $= 2{\cdot}718$; log e $= 0{\cdot}4343$; $\pi = 3{\cdot}142$

$$\frac{e^2}{4\pi\varepsilon_0} = 2{\cdot}307 \times 10^{-28} \text{ m C}^2 \text{ F}^{-1}$$